BARRON'S

Regents Exams and Answers

Geometry
Sixth Edition

Andre Castagna, Ph.D.

Published by Kaplan North America, LLC, d/b/a Barron's Educational Series
1515 West Cypress Creek Road
Fort Lauderdale, Florida 33309
www.barronseduc.com

ISBN: 978-1-5062-9649-4

10 9 8 7 6 5 4 3 2 1

Kaplan North America, LLC, d/b/a Barron's Educational Series print books are available at
special quantity discounts to use for sales promotions, employee premiums, or educational
purposes. For more information or to purchase books, please call the Simon & Schuster
special sales department at 866-506-1949.

Dedication

To my loving wife, Loretta, who helped make this endeavor possible with her unwavering support, and my geometry buddies Eva, Rose, and Henry.

Contents

Regents Exams, Answers, and Self-Analysis Charts   227

How to Use This Book

This book is a valuable tool to aid in your Regents exam preparation. Begin with the Overview of the Geometry Regents Exam in the next section to understand the layout and topics covered in the exam, how points are awarded, and how your score is determined. The Test-Taking Tips and Strategies section provides critical information on the tools needed, use of the reference sheet, and general strategies for exam preparation.

As you begin your review, make use of the section Key Geometry Topics: A Quick Review and Practice to help reinforce your understanding of all the topics and theorems covered in the exam. In particular, note the important theorems, formulas, and definitions that you need to memorize. Anything covered in this section that does not appear on the reference sheet should be memorized.

As you move on to the practice exams, use the self-analysis charts to help identify areas where extra study and practice are needed. Make use of the detailed solutions and go back to the review section to help improve your understanding in any areas where you missed points.

Teachers can use this book as a classroom aid. The content is fully up to date with the Next Generation Learning Standards. Topics and practice problems are organized sequentially so any required topics for a section have been covered in previous sections. The review section can be used for self-paced study and pairs well with *Let's Review Regents Geometry*.

Overview of the Regents Geometry Exam

FORMAT OF THE GEOMETRY REGENTS EXAM

The questions on the Next Generation Learning Standards Geometry Regents can be divided into 6 broad topic areas. Some topics show up more frequently than others, such as transformations, congruence, similarity, trigonometry, and coordinate geometry.

Topic Area	Percent of Exam
Congruence (transformations, triangles, parallelograms)	27%–34%
Similarity, right triangles, and trigonometry	29%–37%
Circles	2%–8%
Coordinate geometry	12%–18%
Volume and cross sections	2%–8%
Modeling with geometry	8%–15%

The exam is 3 hours long and consists of four parts as shown in the table below.

Part	Number of Questions	Type of Question	Points per Question	Total Number of Points
I	24	Multiple-choice	2	48
II	7	Constructed-response	2	14
III	3	Constructed-response	4	12
IV	1	Constructed-response	6	6
Total	**35**	—	—	**80**

HOW THE TEST IS SCORED

Part I is multiple-choice. You do not need to show work to get credit, but no partial credit can be earned. Parts II through IV are constructed-response. Correct work must be shown to get full credit, and partial credit is available according to the scoring rubric issued by the New York State Department of Education.

When determining partial credit, the rubrics often distinguish between *computational errors* and *conceptual errors*. A computational error might be an error in your algebra, graphing, or rounding. Computational errors will generally cost you 1 point, no matter how many points the question is worth. Examples of conceptual errors include using the incorrect formula (volume of a cone instead of a prism) or applying an incorrect relationship (congruent instead of supplementary alternate interior angles). Half the credit of the problem is generally deducted for conceptual errors.

Because partial credit is awarded on the constructed-response questions, it is extremely important to show all your work. A correct answer with no work shown will usually cost you most of the points available for that problem.

Your total number of points earned for all four parts will be added to determine your raw score. A conversion table specific for each exam is then used to convert your raw score to a final scaled score, which is reported to your school. The actual conversion charts are included at the end of each of the Regents exams in this book so you can convert your raw score to a final scaled score. The percentage of points needed to earn a final score of 65% is usually less than 65%, but the effect of the "curve" diminishes as your raw score increases.

Test-Taking Tips and Strategies

CALCULATOR, COMPASS, STRAIGHTEDGE, PEN, AND PENCIL

Graphing calculators are required for Next Generation Learning Standards Geometry Regents exam, and schools must provide one to any student who doesn't have their own. You may bring your own calculator if you own one, and many students feel more comfortable using their own familiar calculator. Be aware, though, that test administrators may clear the memory and any stored programs you might have saved on your calculator before the exam begins.

Any calculator provided to you will likely have had its memory cleared as well. This process will restore the calculator to its default settings, which may not be the ones you are familiar with and have used in the past. The most common setting that may affect your work is the degree mode versus radian mode. Some calculators have radian mode as the default. It is a good idea to find out what calculator will be provided to you and how to switch between degree and radian modes.

A good working knowledge of the graphing calculator will let you use more than one method to solve a problem. The following calculator techniques are worth knowing:

- Using the graph and intersect features to solve linear and quadratic equations
- Making tables to find patterns or applying trial and error to solve a problem quickly
- Finding the square root and cube root of a number
- Using the trigonometric and inverse trigonometric functions

You will also be provided with a compass and straightedge if you do not have your own. Bringing your own compass is highly recommended since the compasses provided by your school are of unknown quality. Also, it will be less stressful on test day to work with a compass that you are familiar with.

The straightedge provided to you should be used only for making straight lines when graphing or doing constructions. The straightedge may have length markings in inches or centimeters, but these should not be used for determining the length of a segment or checking if two segments are congruent.

You are responsible for bringing your own pen and pencil to the exam. All work must be done in blue or black pen. The only exceptions are graphs and diagrams, which may be done in pencil.

SCRAP PAPER AND GRAPH PAPER

You are not allowed to bring your own scrap paper or graph paper into the exam. The exam booklet contains one page of scrap paper and one page of graph paper. These are perforated and can be removed from the booklet to make working with them easier. Any work you put on these sheets is *not* graded. If you want a grader to consider any work there, you must copy it into the appropriate space in the booklet.

REFERENCE SHEET

The following reference sheet is provided to you during the Regents exam. The same reference sheet is used for the Algebra, Geometry, and Algebra II exams. The three sequence and series formulas and the exponential growth and decay formulas are not part of the Next Generation Learning Standards Geometry curriculum, so do not be concerned with them. You should be familiar with all the other formulas. Remember that anytime you are asked to calculate an area or volume, check the reference sheet. Many of those formulas are provided.

Next Generation High School Math Reference Sheet
(Algebra I, Geometry, Algebra II)

Conversions

1 inch = 2.54 centimeters	1 kilometer = 0.62 mile	1 cup = 8 fluid ounces
1 meter = 39.37 inches	1 pound = 16 ounces	1 pint = 2 cups
1 mile = 5280 feet	1 pound = 0.454 kilogram	1 quart = 2 pints
1 mile = 1760 yards	1 kilogram = 2.2 pounds	1 gallon = 4 quarts
1 mile = 1.609 kilometers	1 ton = 2000 pounds	1 gallon = 3.785 liters
		1 liter = 0.264 gallon
		1 liter = 1000 cubic centimeters

Formulas

Triangle $A = \frac{1}{2}bh$	Pythagorean Theorem $a^2 + b^2 = c^2$
Parallelogram $A = bh$	Quadratic Formula $x = \dfrac{-b \pm \sqrt{b^2 - 4ac}}{2a}$
Circle $A = \pi r^2$	Arithmetic Sequence $a_n = a_1 + (n-1)d$
Circle $C = \pi d$ or $C = 2\pi r$	Geometric Sequence $a_n = a_1 r^{n-1}$
General Prism $V = Bh$	Geometric Series $S_n = \dfrac{a_1 - a_1 r^n}{1 - r}$ where $r \neq 1$
Cylinder $V = \pi r^2 h$	Radians 1 radian $= \frac{180}{\pi}$ degrees
Sphere $V = \frac{4}{3}\pi r^3$	Degrees 1 degree $= \frac{\pi}{180}$ radians
Cone $V = \frac{1}{3}\pi r^2 h$	Exponential Growth/Decay $A = A_0 e^{k(t-t_0)} + B_0$
Pyramid $V = \frac{1}{3}Bh$	

PREPARING FOR THE EXAM

Don't try to cram for the Geometry Regents the day before the exam, or even the week before. Successful students begin preparing months ahead of time. Make the most of your class time during the school year. Take good notes, attempt all homework and classwork, and—most importantly—ask questions when you encounter something you don't understand. A great resource is to save all your exams and quizzes in a folder or binder. They are excellent for review and can help you document your progress.

When taking practice exams, use the self-analysis charts at the end of each Regents exam in this book to record the number of points you earn within each topic area. Then use those results to help identify areas for additional practice. It is a good idea to work through at least one Regents exam under "exam conditions"—no distractions, no breaks, and a 3-hour time limit. Measure how much time you spend on each part, and use that as a guide to help establish a time-management plan for the actual exam.

Note any vocabulary words you are unfamiliar with. Making flash cards is also a great way to improve your vocabulary. Use the glossary in this book to help with understanding geometry vocabulary. You also want to be familiar with the formula sheet. You should know what is on the formula sheet, what needs to be memorized, and what each of the symbols and variables represents.

Test-Taking Strategies:

Here are some good strategies that can be applied to many problems:

1. Sketch a figure if none is provided. Mark up figures with relevant dimensions, congruence markings, parallel markings, and so on.

2. Try to break down complex, multistep problems into smaller pieces. Look for relationships that you can apply.

3. For proofs, decide if you are going to write a two-column or a paragraph proof.

4. If you don't see how to get to a final answer at first, try applying any formulas or relationships that you recognize. Sometimes finding a certain length or angle may help you see the path to the final answer. Partial credit is often earned this way.

5. Don't spend too much time on any one problem. If you are having difficulty with a problem, circle it and come back to it at the end. The exam will contain a mix of easy, moderate, and difficult questions. You don't want to run out of time before doing all of the easy problems.

6. Be sure to express the answer in the requested form. This may include rounding, writing an answer in terms of π, or simplifying radicals.

7. If your answer to a multiple-choice question is not one of the choices, check if your answer can be written another way. For example, try simplifying fractions or radicals, or try rearranging equations.

8. Do the following to maximize your partial credit:
   - Show all work for Parts II, III, and IV. Most of the credit is lost if you write a correct answer with no supporting work.
   - Write any formulas used in Parts II, III, and IV, and then show each step of the evaluation.
   - Work through the entire problem even if you are not sure if you are correct. If you make an error along the way but the work that follows is consistent with the error, you will likely receive partial credit for using the correct method.
   - Do not leave any questions blank. There is no penalty for guessing.

9. Consider alternative methods to solving a problem:
   - Trial and error is a valid method for solving an equation or problem. However, you must clearly write the equation you are using along with at least *three* guesses with appropriate checks. Even if your first guess is correct, demonstrate that you understand the method by showing work for two incorrect guesses and their checks.
   - Some problems can be solved graphically even though a coordinate plane grid is not provided with the problem. You can use the provided scrap graph paper in these cases. The scrap graph paper is not graded. However, this is not a concern for multiple-choice problems since your work does not need to be shown. For some constructed-response questions, sketching the problem and solution on the scrap graph paper may be good way to check your result.

10. Eliminate obviously incorrect choices on multiple-choice questions. For example, you may be able to rule out choices if you expect an obtuse or acute angle for an answer or expect a length to be longer or shorter than others. Use the fact that unless otherwise specified, figures are approximately to scale.

11. Make your paper easy to grade. Your work should be neat and flow from top to bottom. Cross out any incorrect work you do not want the graders to consider on constructed-response questions. You will lose credit if you show both a correct and an incorrect answer. Follow directions if you need to change an answer on the bubble sheet when answering multiple-choice questions. Ask a proctor if you are unsure.

Check your work! If you finish the exam with time left, go back and check as much of your work as you can. After taking the time to prepare for the exam, it would be a waste of that effort to not make use of any extra time you might have.

On Test Day

Be mentally prepared by getting a good night's sleep. Don't try to cram for the exam the night before. Remember that you can retake the exam if the grade you earn does not meet the expectation you set for yourself. So don't panic. Give yourself plenty of time to get to the test. Late students may be admitted to the exam but will not be given additional time. Check with your teachers for your school's policies.

Come prepared with all materials—student ID, pens, pencils, eraser, compass, ruler, and calculator (if you have one). Do *not* bring any restricted items such as cell phones or other electronic devices. Ask ahead of time what items are prohibited.

Most students find that 3 hours is more than enough time for the exam. If you finish early, make use of the extra time by checking your work or re-working some of the problems.

You have worked hard, so be confident. Good luck on the Geometry Regents exam!

Key Geometry Topics: A Quick Review and Practice

3.1 ANGLE, LINE, AND PLANE RELATIONSHIPS

NOTATION

Points, segments, lines, and planes are represented using the following notation:

Name	Definition	Figure	Notation
Point	A location in space with no length, width, or thickness	$\bullet$ A	Point A
Line	An infinitely long set of points that has no width or thickness	A B M N O m	$\overleftrightarrow{AB}$ $\overleftrightarrow{MN}, \overleftrightarrow{NO},$ or $\overleftrightarrow{MO}$ Line m
Ray	A portion of a line with one endpoint and including all points on one side of the endpoint	A B	$\overrightarrow{AB}$

Segment	A portion of a line bounded by two endpoints		$\overline{FG}$
Length of a segment	The distance between the two endpoints of a segment		FG
Plane	A flat set of points with no thickness that extends infinitely in all directions		Plane ABC Plane R
Angle	A figure formed by two rays with a common endpoint		$\angle RST$ or $\angle S$
Angle measure	The amount of opening of an angle, measured in degrees or radians		$\angle 1$ $m\angle 1$

An angle can be classified as a

- **Right angle**—measures 90°
- **Acute angle**—measures less than 90°
- **Obtuse angle**—measures greater than 90° and less than 180°
- **Straight angle**—measures 180°

INTERSECTING, PARALLEL, PERPENDICULAR, AND SKEW

- The intersection of two lines is one point (lines r and s intersecting at point M).
- The intersection of two planes is one line (planes ABC and ABD intersect along $\overleftrightarrow{AB}$).

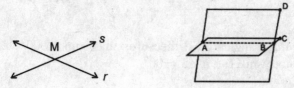

- *Coplanar lines* lie in the same plane; they are either parallel ($\overleftrightarrow{AD}$ and $\overleftrightarrow{BC}$) or intersect ($\overleftrightarrow{AD}$ and $\overleftrightarrow{AB}$).
- *Parallel lines* are coplanar and never intersect. The symbol for parallel is $\parallel$ ($\overleftrightarrow{AD} \parallel \overleftrightarrow{BC}$).
- *Skew lines* are lines that are not coplanar. $\overleftrightarrow{AE}$ and $\overleftrightarrow{GF}$ are skew.
- *Perpendicular lines* intersect at right angles, which measure 90°. The symbol for perpendicular is $\perp$ ($r \perp s$). Line $r \perp$ line s in the accompanying figure. The small square at the point of intersection is a symbol for a right angle. $m\angle ABC + m\angle DEF = 180°$.

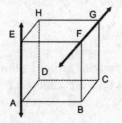

BASIC ANGLE RELATIONSHIPS

- *Supplementary* angles have measures that sum to 180°. ∠ABC and ∠DEF are supplementary.

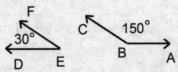

- *Complementary* angles have measures that sum to 90°. ∠1 and ∠2 are complementary. m∠1 + m∠2 = 90°.

- The sum of the measures of adjacent angles around a point equals 360°. m∠1 + m∠2 + m∠3 + m∠4 = 360°.

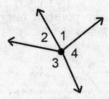

- The sum of the measures of adjacent angles around a line equals 180°. Two adjacent angles that form a straight line are called a *linear pair*, and are supplementary. m∠1 + m∠2 = 180°. ∠1 and ∠2 are a linear pair.

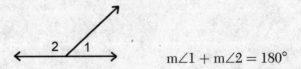

$$m\angle 1 + m\angle 2 = 180°$$

- The measure of a whole equals the sum of the measures of its parts. m∠ABC = m∠1 + m∠2 + m∠3.

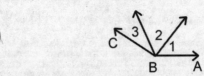

- *Vertical angles* are the congruent opposite angles formed by intersecting lines. ∠1 and ∠3 are vertical angles; therefore, ∠1 ≅ ∠3. ∠2 and ∠4 are vertical angles; therefore, ∠2 ≅ ∠4.

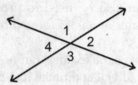

- *Angle bisectors* divide angles into two congruent angles. $\overrightarrow{CD}$ bisects ∠ACB, ∠ACD ≅ ∠BCD.

ANGLES FORMED BY PARALLEL LINES

When two parallel lines (lines *m* and *n*) are intersected by a transversal (line *v*), eight angles are formed. Any pair of these angles will be either supplementary or congruent. Certain pairs are given special names. The named pairs have the following properties:

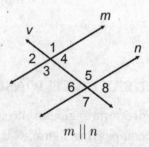

$$m \parallel n$$

Alternate interior angles are congruent. Alternate interior angles lie between parallel lines *m* and *n* on opposite sides of transversal *v*.

Same side interior angles are supplementary. Same side interior angles lie between parallel lines *m* and *n* on the same side of transversal *v*. ∠4 and ∠5 are supplementary, ∠3 and ∠6 are supplementary. m∠4 + m∠5 = 180°, m∠3 + m∠6 = 180°.

Corresponding angles are congruent. Corresponding angles are the interior/exterior pair of congruent angles on the same side of the transversal. $\angle 4 \cong \angle 8$, $\angle 1 \cong \angle 5$, $\angle 2 \cong \angle 6$, $\angle 3 \cong \angle 7$.

These relationships can also be used to prove two lines are parallel. For example, if the alternate interior angles formed by two lines are congruent, then the lines are parallel.

If a problem involves two parallel lines, but no transversal that intersects both lines, it may be helpful to add an additional line called an auxiliary line. In order to find m$\angle BED$ in the accompanying figure, the auxiliary line $\overleftrightarrow{FEG}$ is drawn parallel to $\overline{CD}$ and $\overline{AB}$. Now the two alternate interior angle relationships can be used:

$$m\angle FED = m\angle CDE = 38° \quad \text{alternate interior angles}$$
$$m\angle FEB = m\angle ABE = 32° \quad \text{alternate interior angles}$$
$$m\angle DEB = m\angle FED + m\angle FEB \quad \text{angle addition}$$
$$= 38° + 32°$$
$$= 70°$$

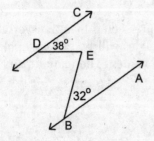 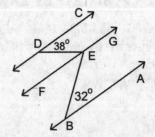

BASIC LINE/SEGMENT/RAY RELATIONSHIPS

In segments, the point that divides a segment into two congruent segments is called a *midpoint*. In the accompanying figure, S is the midpoint of $\overline{RT}$, and $\overline{RS} \cong \overline{ST}$. Any line, segment, or ray that intersects a segment at its midpoint is called a *segment bisector*, or simply *bisector*. Line m bisects $\overline{RT}$ at its midpoint S. Lines can bisect segments, but lines cannot be bisected themselves because they do not have a finite length.

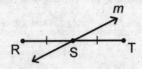

- A line, segment, or ray that is perpendicular to and bisects a segment is called a perpendicular bisector. $\overleftrightarrow{CD}$ is the perpendicular bisector of $\overline{AB}$. The four angles formed by their intersection are right angles, and $\overline{AE} \cong \overline{BE}$.

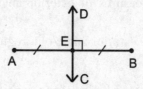

As with angles, the length of a divided segment is equal to the sum of its parts. In the accompanying figure, $MN + NO = MO$. If point N is also a midpoint, then the additional relationship $MN = NO$ would be true.

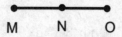

ANGLES IN POLYGONS

Definitions:

- **Polygon**—a closed planar figure with straight sides
- **Regular polygon**—a polygon whose sides and angles are all congruent
- **Interior angle**—the angle inside the polygon formed by two adjacent sides
- **Exterior angle**—the angle formed by a side and the extension of an adjacent side in a polygon
- **Triangle**—a polygon with 3 sides
- **Quadrilateral**—a polygon with 4 sides
- **Pentagon, hexagon, octagon, decagon**—polygons with 5, 6, 8, and 10 sides

In the accompanying pentagon, angles 1 through 5 are interior angles and angles 6 through 10 are exterior angles.

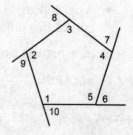

The following relationships can be used to find interior and exterior angles of a polygon:

- The sum of the measures of interior angles of a polygon with n sides equals $180° \times (n - 2)$.

- The measure of one interior angle of a regular polygon with n sides equals $\dfrac{180°(n-2)}{n}$.

- The sum of the measures of exterior angles of a polygon with n sides equals $360°$.

- The measure of one exterior angle of a regular polygon with n sides equals $\dfrac{360°}{n}$.

Using the accompanying figure of the pentagon on page 14:

$$m\angle 1 + m\angle 2 + m\angle 3 + m\angle 4 + m\angle 5 = 180 \times (n-2)$$
$$= 180(5-2)$$
$$= 540°$$
$$m\angle 6 + m\angle 7 + m\angle 8 + m\angle 9 + m\angle 10 = 360°$$

If the pentagon is regular,

$$m\angle 1 = \frac{180°(n-2)}{n}$$
$$= \frac{180°(5-2)}{5}$$
$$= 108°$$
$$m\angle 6 = \frac{360°}{n}$$
$$= \frac{360°}{5}$$
$$= 72°$$

- A central angle is the angle formed by connecting the center of the polygon to two adjacent vertices, as shown. The measure of a central angle in a regular polygon with n sides equals $\dfrac{360°}{n}$. For pentagon $ABCDE$, $m\angle APB = \dfrac{360°}{5} = 72°$.

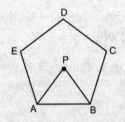

Practice Exercises

1 In the figure of the rectangular prism, which of the following is true?

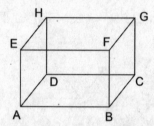

  (1) Points E, H, D, and A are coplanar and collinear.

  (2) $\overline{HD}$ is skew to $\overline{CD}$, and $\overline{CD} \perp \overline{CG}$.

  (3) $\overline{EA} \parallel \overline{CG}$, and $\overline{EH}$ is skew to $\overline{FB}$.

  (4) $\overline{EA} \perp \overline{BC}$, and $\overline{AB} \parallel \overline{CD}$.

2 Which of the following would be the best definition of an angle?

  (1) the union of two rays with a common endpoint

  (2) a geometric figure measured in degrees

  (3) one-third of a triangle

  (4) a line that bends

3 $\overleftrightarrow{BFA} \perp \overrightarrow{CF}$ and m∠CFE = 42°. Find m∠BFD.

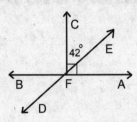

(1) 42°   (3) 48°

(2) 45°   (4) 52°

4 $\overrightarrow{BC}$ bisects ∠ABD. If m∠ABD = $(8x - 12)°$ and m∠ABC = $(3x + 4)°$, find m∠ABD.

(1) 10°   (3) 34°

(2) 20°   (4) 68°

5 Line p intersects lines m and n. For what values of x could you conclude that m is parallel to n?

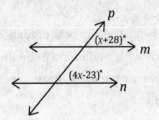

(1) 12°   (3) 35°

(2) 17°   (4) 62°

6 Points F, G, H, and I are collinear. G is the midpoint of $\overline{FH}$, H is the midpoint of $\overline{GI}$, $FG = (x^2 + 5x)$, and $\overline{HI} = 3x + 8$.

What is the length of $\overline{FG}$?

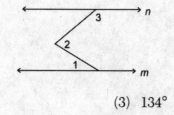

        F   G   H   I

(1) 14   (3) 24

(2) 20   (4) 32

7 In the accompanying figure, lines m and n are parallel. If m∠1 = 32° and m∠2 = 78°, what is the measure of ∠3?

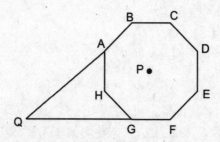

(1) 102°   (3) 134°

(2) 110°   (4) 148°

8 A regular polygon has interior angles that each measure 156°. How many sides does the polygon have?

9 Regular octagon $ABCDEFGH$ is shown in the figure below.

Sides $\overline{AB}$ and $\overline{FG}$ are both extended to Q.

What is the measure of ∠Q?

(1) 22.5°   (3) 45°

(2) 30°   (4) 60°

10 In regular hexagon $ABCDEF$, $\overline{AD}$ and $\overline{FC}$ intersect at O. What is the measure of $\angle AOF$?

11 In the accompanying figure, $\overline{WPX}$ intersects $\overline{YPZ}$ at P.

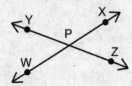

Prove the theorem that the opposite angles formed by intersecting lines are congruent, $\angle WPY \cong \angle ZPX$.

Solutions

1 Choice (1) is not correct since E, H, and D do not lie on a single straight line and are not collinear.

Choice (2) is not correct since $\overline{HD}$ and $\overline{CD}$ are perpendicular, so they cannot be skew.

Choice (4) is not correct since $\overline{EA}$ and $\overline{BC}$ do not intersect, so they cannot be perpendicular.

The correct choice is **(3)**.

2 Considering each choice:

(1) This definition is precise and uses previously defined geometric terms.

(2) This definition is not precise. A counterexample would be an arc, which can also have a degree measure.

(3) Not all angles are part of triangles.

(4) This definition is not precise, since "bend" is not a defined geometric term.

The correct choice is **(1)**.

3 $\text{m}\angle CFE + \text{m}\angle EFA = 90°$ Perpendicular lines form 90° angles.

$\phantom{\text{m}\angle CFE}42° + \text{m}\angle EFA = 90°$

$\phantom{\text{m}\angle CFE + 42° }\text{m}\angle EFA = 48°$

$\phantom{\text{m}\angle CFE + }\text{m}\angle BFD = \text{m}\angle EFA$ Vertical angles are congruent.

$\phantom{\text{m}\angle CFE + }\text{m}\angle BFD = 48°$

The correct choice is **(3)**.

4 m∠ABD = m∠ABC + m∠CBD   angle addition

   m∠ABC = m∠CBD   bisector forms

                                        2 ≅ segments

   m∠ABC + m∠ABC = m∠ABD   substitute m∠ABC for

                                            m∠CBD

$$3x + 4 + 3x + 4 = 8x - 12$$
$$6x + 8 = 8x - 12$$
$$x = 10$$

$$\text{m}\angle ABD = (8 \cdot 10 - 12)° \quad \text{evaluate m}\angle ABD \text{ for}$$
$$x = 10$$

$$= 68°$$

The correct choice is **(4)**.

5 The two angles indicated in the figure are corresponding angles, so lines m and n are parallel if the corresponding angles are congruent, or have the same angle measure.

$$4x - 23 = x + 28$$
$$3x = 51$$
$$x = 17$$

The correct choice is **(2)**.

6 A bisector divides a segment into two congruent halves, so $FG = GH$ and $GH = HI$. Combine these relationships to obtain $FG = HI$.

$$x^2 + 5x = 3x + 8$$
$$x^2 + 2x - 8 = 0$$

$(x - 2)(x + 4) = 0$		factor
$(x - 2) = 0$	$(x + 4) = 0$	zero product property
$x = 2$	$x = -4$	

Eliminate the solution $x = -4$, because that would lead to a negative or zero length; therefore, $x = 2$.

$$FG = x^2 + 5x$$
$$= 2^2 + 5 \cdot 2$$
$$= 14$$

The correct choice is **(1)**.

7 Sketch the auxiliary line parallel to m and n as shown.

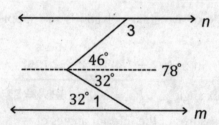

The 78° angle is divided into a lower and upper part. The lower part is an alternate interior angle to $\angle 1$, so it must measure 32°. The measure of the upper part is equal to $78° - 32° = 46°$.

$\angle 3$ and the 46° angle are same side interior angles, so they must be supplementary.

$$m\angle 3 + 46° = 180°$$
$$m\angle 3 = 134°$$

The correct choice is **(3)**.

8 Each interior angle of a polygon measures $\dfrac{180°(n-2)}{n}$ degrees.

$$\frac{180°(n-2)}{n} = 156$$
$$180(n-2) = 156n$$
$$180n - 360 = 156n$$
$$24n = 360$$
$$n = 15$$

The polygon has 15 sides.

9 The sum of the exterior angles of a polygon equals 360°, and in a regular polygon each exterior angle measures $\dfrac{360°}{n}$.

The strategy is to find each interior angle of quadrilateral $QAHG$.

$$m\angle QAH = m\angle QGH = \frac{360°}{8}$$
$$= 45°$$

Each interior angle of a regular polygon is the supplement of an exterior angle.

$$m\angle AHG = 180° - 45° = 135°$$

The exterior convex angle $\angle AHG = 360° - 135° = 225°$.

The sum of the measures of the interior angles of

$$QAHG = (n-2) \cdot 180°$$
$$= (4-2) \cdot 180°$$
$$= 360°$$

$$m\angle Q + m\angle QAH + m\angle AHG + m\angle QGH = 360°$$
$$m\angle Q + 45° + 225° + 45° = 360°$$
$$m\angle Q + 315° = 360°$$
$$m\angle Q = 45°$$

The correct choice is **(3)**.

10 The intersection of the two diagonals $\overline{FC}$ and $\overline{AD}$ is the center of the regular polygon, so $\angle AOF$ is a central angle.

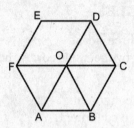

$$\text{m}\angle AOF = \frac{360°}{n} \quad \text{central angle formula for a regular polygon}$$
$$= \frac{360°}{6}$$
$$= 60°$$

11 The strategy is to show $\angle WPY$ and $\angle ZPX$ are supplementary to the same angle.

Statement	Reason
1. $\overleftrightarrow{WPX}$ intersects $\overleftrightarrow{YPZ}$ at P.	1. Given
2. $\angle WPY$ and $\angle YPX$ are a linear pair. $\angle ZPX$ and $\angle YPX$ are a linear pair.	2. Definition of a linear pair
3. $\angle WPY$ and $\angle YPX$ are supplementary. $\angle ZPX$ and $\angle YPX$ are supplementary.	3. Linear pairs are supplementary
4. $\angle WPY \cong \angle ZPX$.	4. Angles supplementary to the same angle are congruent

3.2 TRIANGLE RELATIONSHIPS

CLASSIFYING TRIANGLES

A triangle is a polygon with three sides.
 Triangles can be classified by their sides or by their angles.

- By Sides
  Scalene—no sides are congruent.
  Isosceles—two sides are congruent.
  Equilateral—three sides are congruent.

- By Angle
  Acute—all angles are acute.
  Right—one angle is a right angle.
  Obtuse—one angle is obtuse.

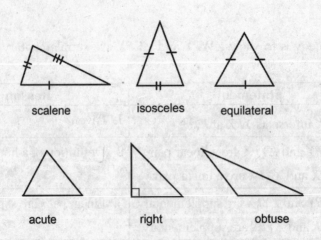

scalene   isosceles   equilateral

acute   right   obtuse

SPECIAL SEGMENTS AND POINTS OF CONCURRENCY IN TRIANGLES

There are four special segments that can be drawn in a triangle, and every triangle has three of each. These are the altitude, median, angle bisector, and perpendicular bisector, shown in the accompanying figure.
 Altitude—a segment from a vertex perpendicular to the opposite side
 Median—a segment from a vertex to the midpoint of the opposite side
 Angle bisector—a line, segment, or ray passing through the vertex of a triangle and bisecting that angle

Perpendicular bisector—a line, segment, or ray that is perpendicular to and passes through the midpoint of a side

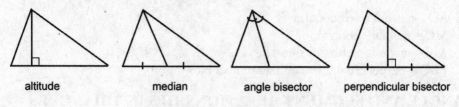

| altitude | median | angle bisector | perpendicular bisector |

Special segments in triangles

The three medians of a triangle will all intersect at a single point called the centroid. We say the medians are *concurrent* at the centroid, or the centroid is the *point of concurrency*. The other special segments will be concurrent as well. The following table illustrates each of the points of concurrency.

Segment	Concurrent at	Feature
Perpendicular bisectors	Circumcenter	Center of the circumscribed circle, equidistant from each of the three vertices
Angle bisectors	Incenter	Center of the inscribed circle, equidistant from the three sides of the triangle
Medians	Centroid	Divides each median in 2 : 1 ratio and is the center of gravity of the triangle
Altitudes	Orthocenter	Located inside acute triangles, on a vertex of right triangles, and outside obtuse triangles

The following phrase can help you remember the points of concurrency:

"**All** **of** **m**y **c**hildren **a**re **b**ringing **in** **p**eanut **b**utter **c**ookies"

AO—altitudes/orthocenter
MC—medians/centroid
ABI—angle bisectors/incenter
PBC—perpendicular bisectors/circumcenter

ANGLE AND SEGMENT RELATIONSHIPS IN TRIANGLES

Angle sum theorem	The sum of the measures of the interior angles of a triangle equals 180°.	 $m\angle A + m\angle B + m\angle C = 180°$
Exterior angle theorem	The measure of any exterior angle of a triangle equals the sum of the measures of the nonadjacent interior angles.	 $m\angle 1 = m\angle 3 + m\angle 4$
Isosceles triangle theorem and its converse	If two sides of a triangle are congruent, then the angles opposite them are congruent. If two angles in a triangle are congruent, then the sides opposite them are congruent.	 $\angle A \cong \angle B, \overline{AC} \cong \overline{BC}$

Equilateral triangle theorem	All interior angles of an equilateral triangle measure 60°.	$$\overline{AB} \cong \overline{BC} \cong \overline{AC},$$ $$m\angle A = m\angle B = m\angle C = 60°$$
Pythagorean theorem	In a right triangle, the sum of the squares of the legs equals the square of the hypotenuse.	$a^2 + b^2 = c^2$
Triangle inequality theorem	In a triangle, the sum of any two side lengths is greater than the length of the third side. Add and subtract any two sides to get the range of possible values of the third side.	$a + b > c$ $b + c > a$ $c + a > b$

Practice Exercises

1 In △FGH, K is the midpoint of $\overline{GH}$. What type of segment is $\overline{FK}$?

  (1) median   (3) angle bisector

  (2) altitude   (4) perpendicular bisector

2 The side lengths of a triangle are 6, 8, and 10.

The triangle can be classified as

  (1) equilateral   (3) obtuse

  (2) acute   (4) right

3 The angle measures of a triangle are $(6x + 6)°$, $(8x − 8)°$, and $(12x)°$. The triangle can be classified as

  (1) obtuse   (3) isosceles

  (2) equilateral   (4) scalene

4 Line m ‖ line n, and △ABC is isosceles with $AC = BC$. What is the measure of ∠3?

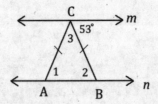

  (1) 53°   (3) 63.5°

  (2) 60°   (4) 74°

5 In △ABC, $\overline{CY}$ is an angle bisector, m∠$AYC = 71°$, and m∠$B = 23°$. What is the measure of angle A?

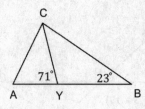

(1) 61°   (3) 76°

(2) 71°   (4) 86°

6 △RST is a right triangle with right angle ∠RST. △PSR is equilateral. Find the measure of angle T.

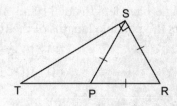

(1) 15°   (3) 30°

(2) 25°   (4) 45°

7 In the figure below, triangle ABC is isosceles with $AC = BC$, m∠$A = 70°$, and $\overleftrightarrow{AC} \parallel \overrightarrow{BE}$. Find the measure of ∠$CBE$.

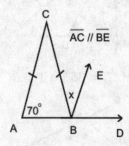

(1) 20°   (3) 35°

(2) 25°   (4) 40°

8 $\overline{JHFK}$ intersects $\overline{IHG}$ at H. If m$\angle KFG = 156°$, m$\angle G = 117°$, m$\angle JHI =$ $(2x + 15)°$, and m$\angle J = (4x + 6)°$, what is the measure of angle J?

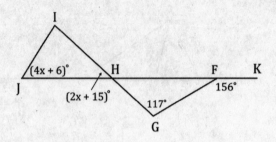

(1) 42°   (3) 62°

(2) 54°   (4) 70°

9 If two sides of a triangle have lengths of 11 and 21, which of the following inequalities represents the possible lengths of the third side?

(1) $11 < x < 22$ (3) $10 < x < 32$

(2) $11 \leq x \leq 22$ (4) $10 \leq x \leq 32$

10 $\angle A$ of $\triangle ABC$ is a right angle, and $\overline{CD}$ is a median. Find the length CD if $AC = 9$ and $BC = 13$.

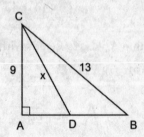

11 In $\triangle ABC$ shown below, m$\angle B = 36°$, $AB = BC$, and $\overline{AD}$ bisects $\angle BAC$.

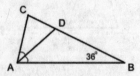

Is $\triangle CAD$ isosceles? Justify your answer.

12 $\overline{RTV}$ intersects $\overline{STU}$ at T.

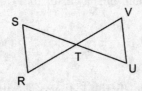

What is the relationship between the sum (m∠R + m∠S) and the sum (m∠V + m∠U)? Explain your reasoning.

Solutions

1 A median is a segment that joins a vertex of a triangle to the midpoint of the opposite side.

   The correct choice is **(1)**.

2 The three side lengths satisfy the Pythagorean theorem:

$$a^2 + b^2 = c^2$$
$$6^2 + 8^2 = 10^2$$
$$36 + 64 = 100$$
$$100 = 100$$

   Therefore, the triangle is a right triangle.

   The correct choice is **(4)**.

3 $6x + 6 + 8x - 8 + 12x = 180$ angle sum theorem
$$26x - 2 = 180$$
$$26x = 182$$
$$x = 7$$

   Substitute $x = 7$ to find the three angle measures.

$$(6 \cdot 7 + 6)^\circ = 48^\circ$$
$$(8 \cdot 7 - 8)^\circ = 48^\circ$$
$$(12 \cdot 7)^\circ = 84^\circ$$

   Since the triangle has two congruent angles, it must have two congruent sides, so it is isosceles.

   The correct choice is **(3)**.

4    $m\angle 2 = 53°$ alternate interior angles
               formed by parallel lines
               are congruent

$m\angle 1 = m\angle 2$ angles opposite congruent
               sides in a triangle are
               congruent

$m\angle 1 = 53°$

$$m\angle 1 + m\angle 2 + m\angle 3 = 180°$$ angle sum theorem
$$53° + 53° + m\angle 3 = 180°$$
$$106° + m\angle 3 = 180°$$
$$m\angle 3 = 74°$$

The correct choice is **(4)**.

5                     $$m\angle AYC = m\angle B + m\angle BCY$$ exterior angle
                                                          theorem in $\triangle YBC$

$$71° = 23° + m\angle BCY$$

$$m\angle BCY = 48°$$

$$m\angle ACY = m\angle BCY$$ angle bisector forms
                             2 congruent angles

$$m\angle ACY = 48°$$

$$m\angle A + m\angle AYC + m\angle ACY = 180°$$ angle sum theorem
                                               in $\triangle AYC$

$$m\angle A + 71° + 48° = 180°$$
$$m\angle A + 119° = 180°$$
$$m\angle A = 61°$$

The correct choice is **(1)**.

6   $m\angle R = m\angle RPS = m\angle PSR = 60°$ equilateral
triangle theorem

     $m\angle RST = 90°$ right angles
measure 90°

     $m\angle T + m\angle RST + m\angle R = 180°$ angle sum theorem
in $\triangle RST$

$$m\angle T + 90° + 60° = 180°$$
$$m\angle T + 150° = 180°$$
$$m\angle T = 30°$$

The correct choice is (**3**).

7   The strategy is to first find $m\angle C$, which must be congruent to $\angle CBE$.

$$m\angle ABC = m\angle A = 70°$$ isosceles triangle
theorem

$$m\angle C + m\angle A + m\angle ABC = 180°$$ triangle angle sum
theorem

$$m\angle C + 70° + 70° = 180°$$
$$m\angle C + 140° = 180°$$
$$m\angle C = 40°$$
$$m\angle CBE = m\angle C$$ congruent alternate
interior angles

$$m\angle CBE = 40°$$

The correct choice is (**4**).

8 Start by finding m$\angle GHK$ by applying the exterior angle theorem in $\triangle FGH$.

$$m\angle GHF + m\angle FGH = m\angle KFG \qquad \text{exterior angle theorem}$$
$$m\angle GHF + 117° = 156°$$
$$m\angle GHF = 39°$$
$$m\angle JHI = m\angle GHF \qquad \text{vertical angles are}$$
$$\text{congruent}$$

$$(2x + 15)° = 39°$$
$$2x = 24$$
$$x = 12$$
$$m\angle J = (4 \cdot 12 + 6)° \qquad \text{substitute } x = 12$$
$$= 54°$$

The correct choice is **(2)**.

9 Subtract and add the two given sides to get 10 and 32. The third side must satisfy the inequality $10 < x < 32$.

The correct choice is **(3)**.

10 Start by using the Pythagorean theorem in $\triangle ABC$ to find AB. From AB, calculate AD, and then use the Pythagorean theorem again on $\triangle ADC$ to find CD.

$$a^2 + b^2 = c^2 \qquad \text{Pythagorean theorem}$$
$$AB^2 + AC^2 = BC^2$$
$$AB^2 + 9^2 = 13^2$$
$$AB^2 + 81 = 169$$
$$AB^2 = 88$$
$$AB = \sqrt{88}$$
$$= \sqrt{4 \cdot 22}$$
$$= 2\sqrt{22}$$

Median $\overline{CD}$ intersects midpoint D, so

$$AD = \frac{1}{2}AB$$
$$AD = \frac{1}{2}\left(2\sqrt{22}\right)$$
$$= \sqrt{22}$$

Now apply the Pythagorean theorem in $\triangle ADC$:

$$a^2 + b^2 = c^2$$
$$AD^2 + AC^2 = CD^2$$
$$\left(\sqrt{22}\right)^2 + 9^2 = CD^2$$
$$22 + 81 = CD^2$$
$$103 = CD^2$$
$$CD = \sqrt{103}$$

11 $\triangle BAC$ has $2 \cong$ sides $\overline{AB}$ and $\overline{BC}$, so it is isosceles. The strategy is to find the measures of $\angle C$, $\angle ADC$, and $\angle CAD$.

$m\angle BAC = m\angle C$	isosceles triangle theorem in $\triangle ABC$
$m\angle BAC + m\angle C + m\angle B = 180°$	angle sum theorem in $\triangle ABC$
$m\angle BAC + \angle BAC + m\angle B = 180°$	substitute $m\angle BAC$ for $m\angle C$
$2m\angle BAC + 36° = 180°$	
$2m\angle BAC = 144°$	
$m\angle BAC = 72°$	
$m\angle C = 72°$	

From the angle bisector, we know

$$m\angle CAD = \frac{1}{2}m\angle BAC$$
$$= \frac{1}{2}(72°)$$
$$= 36°.$$

Find $m\angle CDA$ by applying the angle sum theorem to $\triangle CAD$.

$$m\angle CDA + m\angle C + m\angle CAD = 180°$$
$$m\angle CDA + 72° + 36° = 180°$$
$$m\angle CDA + 108° = 180°$$
$$m\angle CDA = 72°$$

Both $\angle C$ and $\angle CDA$ measure 72°; therefore, $\overline{AC} \cong \overline{AD}$, and $\triangle CAD$ is isosceles.

12 $m\angle RTS + m\angle R + m\angle S = 180°$ from the angle sum theorem. Also, $m\angle VTU + m\angle V + m\angle U = 180°$ for the same reason; therefore,

$$m\angle RTS + m\angle R + m\angle S = m\angle VTU + m\angle V + m\angle U$$

$\angle RTS$ and $\angle VTU$ are congruent vertical angles, so we can subtract their measure from each side of the equation, resulting in

$$m\angle R + m\angle S = m\angle V + m\angle U$$

Therefore, the two sums are equal.

3.3 CONSTRUCTIONS

COPY AN ANGLE

- Given angle $\angle ABC$ and $\overrightarrow{DE}$, construct $\angle EDF$ congruent to $\angle ABC$.
    1. Place point on B and make an arc intersecting the angle at R and S.
    2. With the same compass opening, place point on E and make an arc intersecting at T.
    3. Place point on R and pencil on S and make a small arc.
    4. With the same compass opening, place point on T and make an arc intersecting the previous one at F.
    5. $\angle DEF$ is congruent to $\angle ABC$.

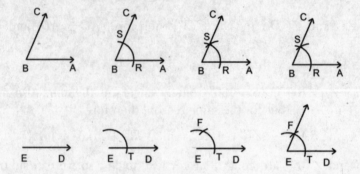

EQUILATERAL TRIANGLE

- Given segment $\overline{AB}$, construct an equilateral triangle with side length AB.
    1. Place point on A and pencil on B and make a quarter circle.
    2. Place point on B and pencil on A and make a quarter circle that intersects the first circle at C.
    3. $\triangle ABC$ is an equilateral triangle.

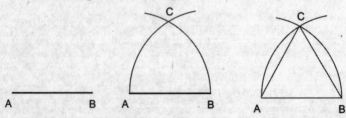

ANGLE BISECTOR

- Given angle ∠ABC, construct angle bisector $\overrightarrow{BD}$.
    1. Place point on B and make an arc intersecting $\overrightarrow{BA}$ and $\overrightarrow{BC}$ at R and S.
    2. Place point on R and make an arc in interior of the angle.
    3. With the same compass opening, place point on S and make an arc that intersects the previous arc at point D.
    4. $\overrightarrow{BD}$ is the angle bisector.

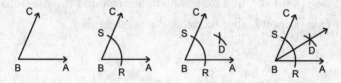

PERPENDICULAR BISECTOR

- Given segment $\overline{AB}$, construct the perpendicular bisector of $\overline{AB}$.
    1. With compass open more than half the length of $\overline{AB}$, place the point at A and make a semicircle running above and below $\overline{AB}$.

       With the same compass opening, place the point at B and make a semicircle running above and below $\overline{AB}$ so that it intersects the first semicircle at R and S.
    2. $\overleftrightarrow{RS}$ is the perpendicular bisector of $\overline{AB}$.

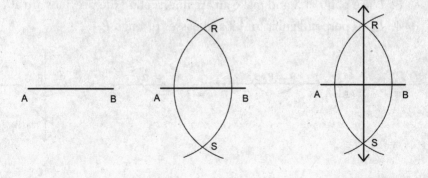

PERPENDICULAR TO LINE FROM A POINT NOT ON THE LINE

- Given line $\overleftrightarrow{AB}$ and point P not on $\overrightarrow{AB}$, construct a line perpendicular to $\overrightarrow{AB}$ passing through P.

    1. Place point at P and make an arc intersecting $\overleftrightarrow{AB}$ at R and S (extend $\overleftrightarrow{AB}$ if necessary).
    2. Place point at R and make an arc on opposite side of line as P.
    3. Place point at S and make an arc intersecting the previous arc at Q.
    4. $\overleftrightarrow{PQ}$ is perpendicular to $\overleftrightarrow{AB}$ and passes through P.

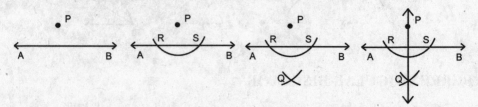

PERPENDICULAR TO LINE FROM A POINT ON THE LINE

- Given line $\overleftrightarrow{AB}$ and point P on $\overrightarrow{AB}$ between A and B, construct a line perpendicular to $\overrightarrow{AB}$ and passing through P.

    1. Place point at P and make an arc intersecting $\overleftrightarrow{AB}$ at R and S.
    2. Place point at R and make an arc.
    3. Place point at S and make an arc intersecting the previous arc at Q.
    4. $\overleftrightarrow{PQ}$ is perpendicular to $\overleftrightarrow{AB}$ and passes through P.

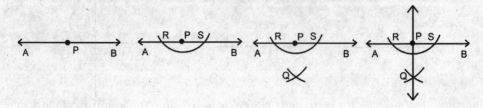

PARALLEL TO LINE THROUGH POINT OFF THE LINE

- Given $\overleftrightarrow{AB}$ and point P not on $\overleftrightarrow{AB}$, construct a line parallel to $\overleftrightarrow{AB}$ and passing through P.

    1. Construct a line passing through P and intersecting $\overleftrightarrow{AB}$ at R.
    2. With point at R, make an arc intersecting $\overrightarrow{PR}$ and $\overleftrightarrow{AB}$ at S and T.
    3. With point at P and the same compass opening, make an arc intersecting $\overrightarrow{PR}$ at U.
    4. With point at T, make an arc intersecting at S.
    5. With point at U and the same compass opening, make an arc intersecting at V.
    6. $\overleftrightarrow{PV}$ is parallel to $\overleftrightarrow{AB}$ and passes through point P.

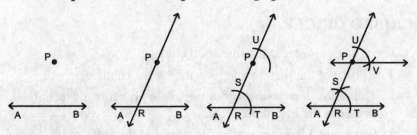

INSCRIBE A REGULAR HEXAGON OR EQUILATERAL TRIANGLE IN A CIRCLE

1. Construct circle P of radius PA.
2. Using the same radius as the circle, place point on A and make an arc intersecting circle at B.
3. Place point at B and make an arc intersecting circle at C. Continue making arcs intersecting at D, E, and F.
4. Connect each point to form regular hexagon $ABCDEF$ or every other point to form equilateral triangle ACE.

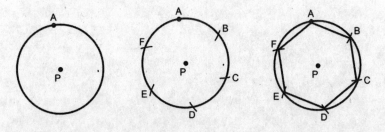

INSCRIBE A SQUARE IN A CIRCLE

- Given circle T
  1. Construct a diameter through the center point T.
  2. Construct the perpendicular bisector of the first diameter, which will also be a diameter.
  3. The intersections of the diameters with the circle are the vertices of the square.

INSCRIBED CIRCLE

- Given $\triangle ABC$, construct the inscribed circle.
  1. Construct the three angle bisectors of the triangle.
  2. The point of concurrency of the angle bisectors, or incenter I, is the center of the inscribed circle.
  3. Construct a line perpendicular to one of the sides of the triangle that passes through the incenter I. Label the point of intersection with the side of the triangle P.
  4. Construct the inscribed circle with center I and radius IP.

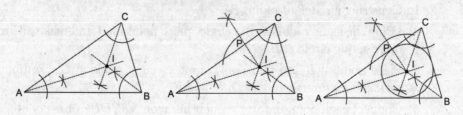

CIRCUMSCRIBED CIRCLE

- Given △ABC, construct the circumscribed circle.

  1. Construct the three perpendicular bisectors of the triangle.
  2. The point of concurrency of the perpendicular bisectors, or circumcenter P, is the center of the circumscribed circle.
  3. Construct the circumscribed circle with center P and radius PA. (The circle should pass through points B and C as well.)

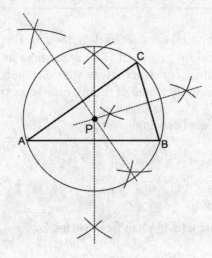

Practice Exercises

Directions: Each of the constructions described in this chapter is a required construction and should be practiced. The constructions in this set of practice problems use the required constructions in various combinations. The first step in each problem is to identify which of the required constructions is called for.

1 Construct a triangle whose angles measure 30°, 60°, and 90°.

2 Given △*JOT*, construct a median from vertex *O*.

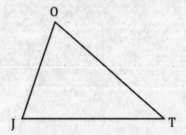

3 Construct the centroid of △*HOT* using a compass and straightedge. Use the compass to confirm your construction by demonstrating the 2 : 1 ratio. Leave all marks of construction.

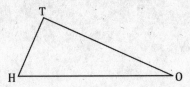

4 Given △*XYZ*, construct the altitude from vertex *X*.

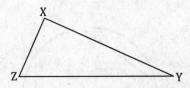

5 Given △*XYZ*, construct a midsegment that intersects sides $\overline{XZ}$ and $\overline{YZ}$.

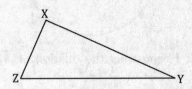

6 Given $\overline{AB}$ and $\overline{BC}$, construct parallelogram *ABCD*.

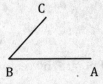

7 Construct a triangle whose angles measure 45°, 45°, and 90° and has $\overline{AB}$ as one of its legs.

A   B

8 Construct a 15° angle.

9 $\overline{JK}$ and $\overline{LM}$ are chords in circle P. Construct the location of point P.

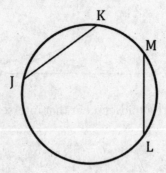

10 Given $\overline{AB}$ and point P, construct the dilation of $\overline{AB}$ with a scale factor of 2 and center at P.

11 Construct the translation of point P by vector $\overrightarrow{AB}$.

12 P' is the reflection of P over line m. Construct line m.

P'

P

13 Construct point Q', the rotation of point Q about center P by $\angle ABC$.

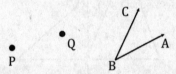

14 Construct a circle inscribed in $\triangle NWK$. [Leave all construction marks.]

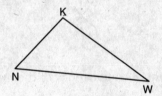

15 Construct a circle circumscribed about $\triangle RBG$. [Leave all construction marks.]

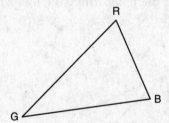

Solutions

1 Construct an equilateral triangle, then bisect one of the angles.

2 Construct A, the midpoint of $\overline{JT}$, using the "perpendicular bisector" construction. Connect O to A. $\overline{OA}$ is a median of $\triangle JOT$.

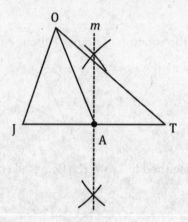

3 The centroid is the point of concurrency of the medians.

- First construct midpoints of two sides of the triangle.
- Then use those midpoints to construct the medians.
- The centroid, G, is the point where the medians intersect.

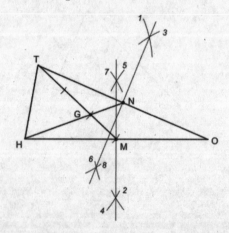

To confirm the centroid divides a median in a 2 : 1 ratio, measure the distance *GM* and show the distance *TG* is twice *GM*:

- Place the compass point at *M* and make an arc at *G*.
- With the same compass opening, place the point at *G* and make a 2nd arc in the direction of *T*.
- With the same compass opening, you should be able to make a 3rd arc starting at the 2nd and ending at *T*.

4 Using the "perpendicular to line from point not on the line" construction, construct a line perpendicular to $\overline{YZ}$ through point *X*. Let *A* be the intersection of the perpendicular line with $\overline{YZ}$. $\overline{AX}$ is the altitude of △*XYZ*.

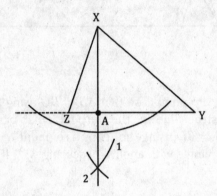

5 Using the "perpendicular bisector" construction, construct the midpoints of $\overline{XZ}$ and $\overline{YZ}$. Label these *A* and *B*. $\overline{AB}$ is a midsegment of △*XYZ*.

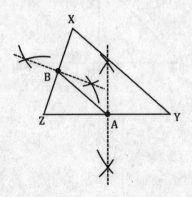

6 The steps are

- Extend $\overline{BA}$.
- Use the "parallel line" construction to construct a line parallel to $\overline{BC}$ and through point A.
- Measure the length BC by making arc 1 with point at B and pencil at C.
- Locate point D by measuring a segment of the same length from point A. D is the fourth vertex of parallelogram $ABCD$.

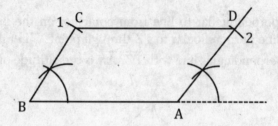

7 Extend $\overline{AB}$, and locate point C so that $AB = BC$ using the "copy a segment" construction. Construct line m, the perpendicular bisector of $\overline{AC}$. Locate point D so that $AB = BD$ using the "copy a segment" construction. $\triangle ABD$ is an isosceles right triangle with angles measuring 45°, 45°, and 90°.

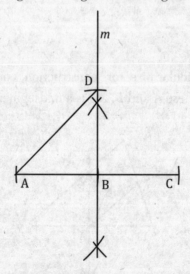

8 Construct an equilateral triangle. Bisect one of its angles to get a pair of 30° angles, and then bisect one of those angles to create two 15° angles.

9 Construct the perpendicular bisectors of $\overline{JK}$ and $\overline{LM}$. The perpendicular bisector of a chord always passes through the center of the circle, so the intersection of the two perpendicular bisectors is point P, the center of the circle.

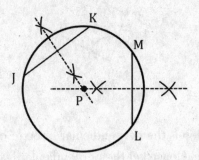

10 The distance from a point to the center is multiplied by the scale factor of the dilation, so $PA' = 2PA$ and $PB' = 2PB$. Construct rays $\overrightarrow{PA}$ and $\overrightarrow{PB}$. With the point at P, measure the distance PA. Move the point to A, and measure to same distance to locate A'. Repeat for point B and B'.

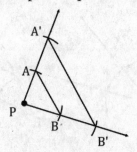

11 Use the "parallel to a line through point off the line" construction to construct a ray through P and parallel to $\overrightarrow{AB}$. With the point at A, measure length AB. Move the point to P, and measure the same distance along the new ray to locate P'. P' is the translation of P by vector $\overrightarrow{AB}$.

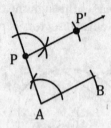

12 The line of reflection is the perpendicular bisector of the segment through a point and its image. Construct the perpendicular bisector of $\overline{PP'}$.

13 Construct a ray through P and Q, and apply the "copy an angle" construction to copy $\angle ABC$ to $\angle PQD$. Then, with the point at P, measure the distance PQ. Use the same distance to locate point Q' along $\overrightarrow{PD}$. Q' is the rotation of Q about point P by $\angle ABC$.

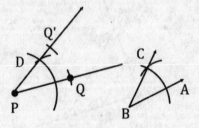

14 Apply the construction for a circle inscribed in a triangle.

15 Apply the construction for a circle circumscribed about a triangle.

3.4 TRANSFORMATIONS

Transformations can be thought of as functions that operate on geometric figures. The original figure is called the preimage and the new figure is called the image. There are four basic transformations to be familiar with—translations, reflections, rotations, and dilations. The first three are rigid motions that do not change the size of the figure and result in a congruent figure. The last one, dilations, can change the size of a figure and results in a similar figure.

TRANSLATIONS

A *translation* slides a figure from one position to another. Translations can be specified by a vector, with the notation $T_{\overrightarrow{FG}}$. A vector consists of a magnitude and a direction. The magnitude is represented by the length of the vector and the direction is indicated by following the endpoint, point F, toward the second point. A translation can also be specified by stating one point and the point it maps to—for example, a translation such that maps E to G. The accompanying figure illustrates $T_{\overrightarrow{FG}}(ABCDE) \rightarrow A'B'C'D'E'$. $ABCDE$ is translated a distance FG parallel to the direction of $\overrightarrow{FG}$. The translation is a rigid motion so $ABCDE \cong A'B'C'D'E'$. Also, segments joining corresponding points are congruent and parallel; $\overline{AA'}$ is congruent and parallel to $\overline{BB'}$, $\overline{CC'}$, $\overline{DD'}$, and $\overline{EE'}$.

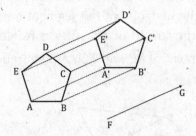

LINE REFLECTIONS

A *line reflection* flips a figure over a line, so that it appears as a mirror image. The line is called the *line of reflection*. The notation r_ℓ indicates a reflection over the line ℓ. The accompanying figure illustrates the reflection of $ABCD$ over line ℓ. Reflections are a rigid motion, so $ABCD \cong A'B'C'D'$. The line of reflection is also the perpendicular bisector of each segment joining corresponding points in the preimage and image. Therefore, every point on the preimage and its corresponding point on the image are equidistant from the line of reflection.

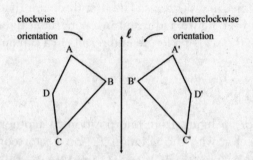

Line reflections do not preserve orientation, which is the direction one must travel to move from one vertex to the next. In the figure, the orientation changes from clockwise to counterclockwise. Noting a change in orientation is one way to identify a reflection.

POINT REFLECTIONS

A *point reflection* maps each point in the preimage to a point such that the center of the reflection is the midpoint of the segment through any corresponding pair of points. $\overline{PQ}$ is the image of $\overline{ST}$ after a reflection through point R. Therefore, R is the midpoint of $\overline{SP}$ and $\overline{TQ}$. A point reflection is equivalent to a 180° rotation.

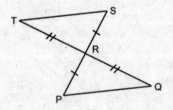

ROTATIONS

A *rotation* is the spinning of a figure about a pivot point called the *center of rotation*. The angle of rotation is measured counterclockwise unless otherwise specified. The accompanying figure shows $ABCD$ rotated about point O to $A'B'C'D'$. The notation $R_{C,a}$ is used to specify a rotation, where a is the angle of rotation and C is the center of rotation. Rotations are rigid motions, so $ABCD \cong A'B'C'D'$. The angle formed by any two corresponding points with the center of rotation as the vertex is equal to the angle of rotation. $m\angle AOA' = m\angle BOB' = m\angle COC' = m\angle DOD' = $ angle of rotation.

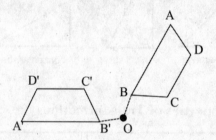

DILATIONS

A *dilation* is a similarity transformation that enlarges or reduces the size of a figure without changing its shape. Angle measures are preserved, and the lengths of segments in the image are proportional to lengths in the preimage. The ratio of lengths is called the scale factor. Dilations are specified about a center point. The distance from a point to the center is also multiplied by the scale factor after a dilation. $D_{C,k}$ is a dilation with a center point C and scale factor k. The accompanying figure shows the dilation $D_{C,2}(\overline{PQ}) \rightarrow \overline{P'Q'}$.

CP' = 2CP, CQ' = 2CQ

P'Q' = 2PQ

HORIZONTAL AND VERTICAL STRETCHES

Two examples of transformations that are neither rigid motions nor similarity transformations are the horizontal stretch and vertical stretch. A *horizontal stretch* will elongate a figure only in the horizontal direction. On the coordinate plane, the *x*-coordinate of every point will be multiplied by a scale factor. A *vertical stretch* will do the same thing to the *y*-coordinate.

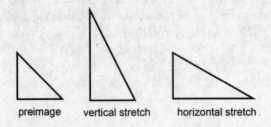

preimage   vertical stretch   horizontal stretch

Summary of the Properties of Transformations

Transformation	What's Preserved	Segments Between Corresponding Points Congruent?	Other Relationships
Reflection $r_{\overleftrightarrow{FE}}(\overline{AB}) \rightarrow \overline{A'B'}$	Corresponding lengths and angles, parallelism	Not necessarily $AA' \neq BB'$	$\overleftrightarrow{FE}$ is the $\perp$ bisector of $\overline{BB'}$ and $\overline{AA'}$ $\overline{AA'} \parallel \overline{BB'}$
Rotation $R_{C,100°}(\overline{AB}) \rightarrow \overline{A'B'}$	Corresponding lengths and angles, parallelism, orientation	Not necessarily $AA' \neq BB'$	$\overline{AC} \cong \overline{A'C}$ $\overline{BC} \cong \overline{B'C}$ $\angle ACA' \cong \angle BCB'$

Transformation	What's Preserved	Segments Between Corresponding Points Congruent?	Other Relationships
Translation $T_{\overline{UV}}(\overline{AB}) \to \overline{A'B'}$	Corresponding lengths and angles, slope, parallelism, orientation	Yes $AA' \neq BB'$	$\overline{AB} \parallel \overline{A'B'}$ $\overline{AA'} \parallel \overline{BB'}$
Dilation $D_{C,3/2}(\overline{AB}) \to \overline{A'B'}$	Corresponding angles, slope, parallelism, orientation	Not necessarily $AA' \neq BB'$	$\overline{AB} \parallel \overline{A'B'}$

Transformations on the Coordinate Plane

The following set of rules can be used to apply a transformation to a point on the coordinate plane.

Reflection	Rotation About the Origin
$r_{x\text{-axis}}(x, y) \to (x, -y)$	$R_{90}(x, y) \to (-y, x)$
$r_{y\text{-axis}}(x, y) \to (-x, y)$	$R_{180}(x, y) \to (-x, -y)$
$r_{y=x}(x, y) \to (y, x)$	$R_{270}(x, y) \to (y, -x)$
$r_{y=-x}(x, y) = (-y, -x)$	

Translation	Dilation
$T_{h,k}(x, y) \rightarrow (x + h, y + k)$	$D_{k,\text{origin}}(x, y) \rightarrow (kx, ky)$

- Rotations of 180° and 270° can also be found by repeating the rule for a 90° rotation.

- Stretches are like dilations in one direction. For horizontal stretches through the origin, multiply only the x-coordinate by the scale factor. For vertical stretches through the origin, multiply only the y-coordinate.

- In a point reflection, the point is the midpoint of the segment between the image and the preimage. You can count jumps from the preimage to the given point and then continue past the given point by the same number of jumps.

COMPOSITIONS

A *composition of transformation* is a sequence of transformations where the image found after the first transformation becomes the preimage for the next transformation. The symbol ° is used to indicate a composition, as in $T_{3,4} \circ D_{3,\text{origin}}$. The transformation on the right is performed first. The same composition can also be expressed as $T_{3,4}(D_{3,\text{origin}})$.

SYMMETRY

A figure has *line symmetry* if a line of reflection can be drawn that divides the figure into two congruent mirror images. For example:

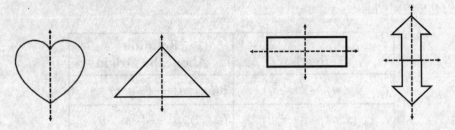

vertical line symmetry   horizontal and vertical line symmetry

A figure has *rotational symmetry* if it can be rotated by some angle $0 < \theta < 360°$ about its center and have every point map to another point on the image. In other words, the image will look identical to the preimage after the rotation.

The accompanying figure shows rectangle $ABCD$ after rotations of 0°, 90°, 180°, 270°, and 360°. Rotations of 0°, 180°, and 360° result in figures that look identical to the original. We say the figure has 180° rotational symmetry—the 0° and 360° rotations are not considered rotational symmetries. Note that the location of individual points has changed. After a 180° rotation, point A is mapped to point C, B is mapped to D, C is mapped to A, and D is mapped to B.

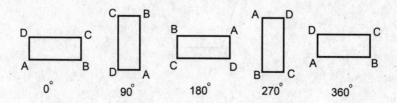

Some figures have more than one rotation that results in an identical figure. An equilateral triangle has 120° and 240° rotational symmetry when rotated about its center.

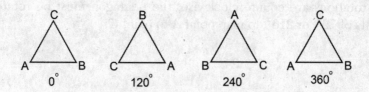

A regular polygon with n sides always has rotational symmetry, with rotations in increments equal to its central angle of $\dfrac{360°}{n}$.

Example

By how many degrees must a regular octagon be rotated so that it maps onto itself? List all the rotations less than 360° that will map an octagon onto itself.

Solution:

$n = 8$ for an octagon.

$$\frac{360°}{n} = \frac{360°}{8} = 45°$$

Any rotation between 0° and 360° that is a multiple of 45° will map the octagon onto itself—45°, 90°, 135°, 180°, 225°, 270°, and 315°.

Example

By how many degrees must pentagon *ABCDE* be rotated about its center to map point *A* to point *C*?

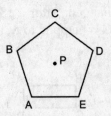

Solution:

First calculate the rotation to map one vertex to the adjacent vertex.

$$\frac{360°}{n} = \frac{360°}{5} = 72°$$

Since rotations are counterclockwise, the pentagon must be rotated three increments of $72°$, or $216°$, to map point *A* to point *C*.

Practice Exercises

1 △*ABC* undergoes a transformation such that its image is △*A'B'C'*. If the side lengths and angles are preserved, but the slopes of the sides are not, the transformation could be a

   (1) reflection or rotation   (3) translation or rotation

   (2) reflection or translation   (4) dilation or translation

2 Which of the following transformations is not a rigid motion?

   (1) reflection   (3) rotation

   (2) dilation   (4) translation

3 Which of the following letters has more than one line of symmetry?

   (1) **A**   (3) **H**

   (2) **C**   (4) **L**

4 Segment $\overline{PQ}$ has endpoints $P(3, 1)$ and $Q(6, 4)$. Find the coordinates of P' and Q' after a horizontal stretch centered at the origin with a scale factor of 3.

   (1) $P'(3, 3), Q'(6, 12)$ (3) $P'(9, 3), Q'(18, 12)$

   (2) $P'(1, 1), Q'(2, 4)$ (4) $P'(9, 1), Q'(18, 4)$

5 Segment $\overline{EF}$ has endpoints $E(-1, 4)$ and $F(2, 5)$. Find the coordinates of E' and F' after a point dilation through the point $Q(3, 6)$.

   (1) $E'(1, 5), F'(3, 6)$ (3) $E'(2, 10), F'(5, 11)$

   (2) $E'(-5, 2), F'(1, 4)$ (4) $E'(7, 8), F'(4, 7)$

6 Which transformation is represented in the figure below?

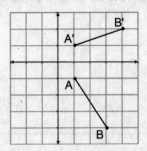

(1) $R_{\text{origin, }90°}$ (3) $r_{x\text{-axis}}$

(2) $R_{\text{origin, }180°}$ (4) $r_{y\text{-axis}}$

7 Which of the following will always map a rectangle onto itself?

   (1) reflection over one of its sides

   (2) reflection over one of its diagonals

   (3) rotation of 180° about its center

   (4) rotation of 90° about one of its vertices

8 The area of a rectangle is 4 cm². The rectangle then undergoes a rotation of 90° about one of its vertices, followed by a dilation with a scale factor of 3, centered at the same vertex as the rotation. What is the area of the image?

   (1) 12 cm²   (3) 32 cm²

   (2) 16 cm²   (4) 36 cm²

9 Regular hexagon *ABCDEF* is shown in the figure below. Which of the following transformations will map the hexagon onto itself such that point *D* maps to point *E*?

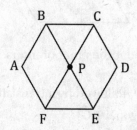

(1) a 60° rotation about point *P* followed by a reflection of line $\overleftrightarrow{BE}$
(2) reflection over line $\overleftrightarrow{BE}$ followed by 60° rotation about point *P*
(3) reflection over line $\overleftrightarrow{CF}$ followed by a translation by vector $\overrightarrow{FE}$
(4) reflection over line $\overleftrightarrow{CF}$ followed by a 120° rotation about point *P*

10 After a certain transformation the image of △*BAD* with coordinates *B*(7, 1), *A*(0, 3), *D*(12, 4) is *B*′(13, 3), *A*′(6, 5), *D*′(18, 1). Is the transformation a translation? Explain why or why not.

11 The vertices of △*MNO* have coordinates *M*(−5, 2), *N*(1, 2), *P*(0, 5). △*M*′*N*′*P*′ is the image of △*MNP* after a translation. Find the coordinates of *N*′ and *P*′ if vertex *M* is mapped to *M*′(3, 6).

12 In the figure below, △*DAB* ≅ △*BED* ≅ △*EBC* ≅ △*CFE*. Which of the following sequences of rigid motions will always map △*DAB* to △*CFE*?

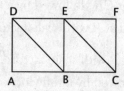

(1) a translation by vector $\overrightarrow{BE}$ followed by a translation by vector $\overrightarrow{EC}$
(2) a rotation of 180° about point *B*, followed by a translation by vector $\overrightarrow{CF}$
(3) a reflection over $\overline{BE}$ followed by a translation by vector $\overrightarrow{BC}$
(4) a translation by vector $\overrightarrow{AC}$ followed by a reflection over $\overline{FC}$

13 Which of the following will map parallelogram $ABCD$ onto itself?

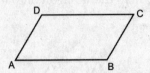

(1) a rotation of 180° about the point of intersection of the two diagonals of $ABCD$

(2) a reflection over $\overline{CD}$, followed by a translation along vector $\overrightarrow{DA}$

(3) a reflection over diagonal $\overline{AC}$

(4) a reflection over $\overline{BC}$, followed by a translation by vector $\overrightarrow{BA}$

14 In the figure below, $\triangle SDR \sim \triangle RCP$. Which of the following compositions of transformations will map $\triangle RCP$ to $\triangle SDR$?

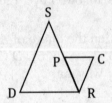

(1) a translation by vector $\overrightarrow{PR}$, followed by a 180° rotation about point R, followed by a dilation centered at R with scale factor $\dfrac{RS}{RP}$

(2) a dilation centered at R with scale factor $\dfrac{RS}{RP}$, followed by a reflection over $\overline{RS}$

(3) a translation by vector $\overrightarrow{CR}$, followed by a reflection over $\overline{PC}$, followed by a dilation centered at R with scale factor $\dfrac{RS}{RP}$

(4) a dilation centered at R with scale factor $\dfrac{RP}{DS}$, followed by a translation along vector $\overrightarrow{PD}$, followed by a reflection over $\overline{DR}$

15 Graph $\triangle A(-2, 1)$, $B(0, 2)$, $C(-1, 4)$. Graph $\triangle A''B''C''$, the image of $\triangle ABC$ after the transformation $T_{1,3} \cdot r_{x=y}$. State the coordinates of $A''B''C''$.

16 Parallelogram *STAR* has coordinates $S(-8, -4)$, $T(-5, -2)$, $A(1, -5)$, and $R(-2, -7)$.

   (a) Graph and state the coordinates of $S'T'A'R'$, the image of *STAR* after $r_{y=-1}$.

   (b) Graph and state the coordinates of $S''T''A''R''$, the image of $S'T'A'R'$ after $T_{7,-2}$.

   (c) Name a single reflection that would map T to T''.

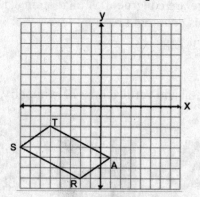

Solutions

1 A translation always preserves slope, so choices (2), (3), and (4) cannot be correct.

 The correct choice is **(1)**.

2 Reflections, rotations, and translations are always rigid motions because the image and preimage are congruent. A dilation may result in an image with a different size, so it is not a rigid motion.

 The correct choice is **(2)**.

3 **H** has both horizontal and vertical lines of symmetry.

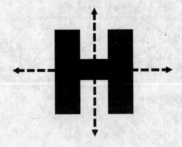

 The correct choice is **(3)**.

4 Multiply each x-coordinate by 3 to get $P'(9, 1)$, $Q'(18, 4)$.

 The correct choice is **(4)**.

5 Count the jumps in both the x- and y-directions from point E to point Q.

$$x: -1 \text{ to } 3 \text{ is an increase of } 4$$
$$y: 4 \text{ to } 6 \text{ is an increase of } 2$$

 Add 4 to the x-coordinate and 2 to the y-coordinate of point E to get $E'(7, 8)$. Follow the same process to find $F'(4, 7)$.

 The correct choice is **(4)**.

6 Approach this problem by simply applying each transformation and checking if $A(1, -1)$ maps to $A'(1, 1)$ and $B(3, -4)$ maps to $B'(4, 3)$. If the correct image results for both, then the choice is correct.

$R_{\text{origin, } 90°}(1, -1) \to (1, 1)$ correct	$R_{\text{origin, } 90°}(3, -4) \to (4, 3)$
$R_{\text{origin, } 180°}(1, -1) \to (-1, 1)$ A' and B' are incorrect	$R_{\text{origin, } 180°}(3, -4) \to (-3, 4)$
$R_{x\text{-axis}}(1, -1) \to (1, 1)$ B' is incorrect	$R_{x\text{-axis}}(3, -4) \to (3, 4)$
$r_{y\text{-axis}}(1, -1) \to (-1, -1)$ A' and B' are incorrect	$r_{y\text{-axis}}(3, -4) \to (-3, -4)$

The correct choice is (**1**).

7 Since a rectangle has twofold symmetry, a rotation of $180°$ about its center will map it to itself.

The correct choice is (**3**).

8 Under a dilation, the area of a polygon is proportional to the scale factor squared, so the area becomes 3^2 times larger, or 9 times larger. The new area is $9 \cdot 4 \text{ cm}^2 = 36 \text{ cm}^2$.

The correct choice is (**4**).

9 A reflection over $\overleftrightarrow{BE}$ will map point D to point F. Each central angle measures $\dfrac{360°}{60°} = 60°$. Therefore, a rotation of $60°$ will map any vertex to the adjacent vertex in the counterclockwise direction. The $60°$ rotation will map a point at vertex F to vertex E.

The correct choice is (**2**).

10 If the transformation is a translation, then every point in the preimage would undergo the same translation $(x, y) \rightarrow (x + h, y + k)$. Calculate each h and k by working backwards from the preimage and image:

$B'(13, 3) = B(7 + h, 1 + k)$	$13 = 7 + h, h = 6$	$3 = 1 + k, k = 2$
$A'(6, 5) = A(0 + h, 3 + k)$	$6 = 0 + h, h = 6$	$5 = 3 + k, k = 2$
$D'(18, 1) = D(12 + h, 4 + k)$	$18 = 12 + h, h = 6$	$1 = 4 + k, k = -3$

The transformation is not a translation because A and B undergo the translation $T_{6,2}$ while D undergoes the translation $T_{6,-3}$.

11 First, find the translation rule.

$T_{h,k} M(-5, 2) \rightarrow M'(-5 + h, 2 + k)$

Using the coordinates $(3, 6)$ for M', solve for h and k.

$-5 + h = 3$	$2 + k = 6$
$h = 8$	$k = 4$

The translation is $T_{8,4}$. Now apply that translation to N and P.

$T_{8,4} N(1, 2) \rightarrow N'(1 + 8, 2 + 4) \rightarrow N'(9, 6)$
$T_{8,4} P(0, 5) \rightarrow P'(0 + 8, 5 + 4) \rightarrow P'(8, 9)$

The correct choice is **(3)**.

12 The figure shows the $\triangle DAB$ after a 180° rotation about B. A translation by $\overrightarrow{CF}$ will then slide the image up and map it to $\triangle CFE$.

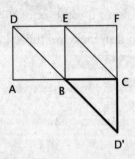

The correct choice is **(2)**.

13 Rotating 180° about the point of intersection of the diagonals will map:
$A \rightarrow C$, $B \rightarrow D$, $C \rightarrow A$, and $D \rightarrow B$.

The other choices will not map *ABCD* onto itself. Choices (2), (3), and (4) result in the following:

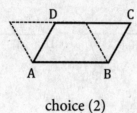

choice (2)

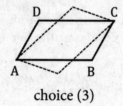

choice (3)

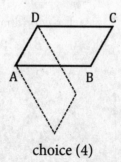

choice (4)

The correct choice is (**1**).

14 The figures show the figure after the translation by $\overrightarrow{PR}$, and then the rotation of $180°$ about R. The final dilation will map C'' to D and R'' to S.

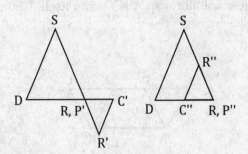

The correct choice is (**1**).

15 In a composition always do the rightmost transformation first. For $T_{1,3} \cdot r_{x=y}$, start with $r_{y=x}$. For a reflection over the line $y = x$, switch the order of the coordinates.

$r_{y=x}$: $A(-2, 1) \rightarrow A'(1, -2)$, $B(0, 2) \rightarrow B'(2, 0)$, $C(-1, 4) \rightarrow C'(4, -1)$

Now use the image from the reflection as the preimage of the translation.

For a translation $T_{1,3}$, add 1 to each x-coordinate and 3 to each y-coordinate.
$T_{1,3}$: $A'(1, -2) \rightarrow A''(2, 1)$, $B'(2, 0) \rightarrow B''(3, 3)$, $C'(4, -1) \rightarrow C''(5, 2)$

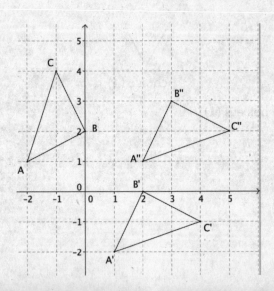

16 a) The x-coordinates of each point in *STAR* are unchanged after a reflection over the line $y = -1$. To find the y-coordinates, count the vertical distance from the point to $y = -1$, then continue by the same distance on the other side of $y = -1$. $S(-8, -4)$ is 3 units below $y = -1$.

The y-coordinate of S' is $-1 + 3 = 2$. $T(-5, -2)$ is 1 unit below $y = -1$.	$S'(-8, 2)$
The y-coordinate of T' is $-1 + 1 = 0$. $A(1, -5)$ is 4 units below $y = -1$.	$T'(-5, 0)$
The y-coordinate of A' is $-1 + 4 = 3$. $R(-2, -7)$ is 6 units below $y = -1$.	$A'(1, 3)$
The y-coordinate of R' is $-1 + 6 = 5$. $S'(-8, 2)\ T'(-5, 0)\ A'(1, 3)\ R'(-2, 5)$	$R'(-2, 5)$

b) Apply the translation $T_{7,-2}$ by adding 7 to each x-coordinate and subtracting 2 from each y-coordinate.

$$S''(-1, 0)\ T''(2, -2)\ A''(8, 1)\ R''(5, 3)$$

c) To reflect $T(-5, -2)$ to $T''(2, -2)$, note that the y-coordinates are the same, so we need a reflection over a vertical line that is halfway between $x = -5$ and $x = 2$.

$$\frac{1}{2}(-5 + 2) = -1\frac{1}{2}$$

A reflection over the line $x = -1\frac{1}{2}$ will map $T(-5, -2)$ to $T''(2, -2)$.

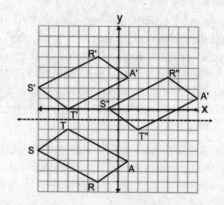

3.5 TRIANGLE CONGRUENCE

CONGRUENCE

Two figures are congruent if one figure can be mapped onto another by a sequence of rigid motions. The congruent figures will have exactly the same size and shape, which means for polygons that all *corresponding* pairs of sides and angles are congruent.

A congruence statement will match up the corresponding pairs of congruent parts.

For example, from the statement $\triangle BUS \cong \triangle CAR$ we can conclude:

$$\angle B \cong \angle C \quad \overline{BU} \cong \overline{CA}$$
$$\angle U \cong \angle A \quad \overline{US} \cong \overline{AR}$$
$$\angle S \cong \angle R \quad \overline{SB} \cong \overline{RC}$$

Once two figures have been proven congruent, we know all pairs of parts are congruent. In triangles, this theorem is abbreviated by CPCTC—corresponding parts of congruent triangles are congruent.

PROVING POLYGONS CONGRUENT BY TRANSFORMATIONS

Polygons can be proven congruent by identifying a sequence of rigid motions that map one onto the other. One method is to show each vertex of the preimage maps to a corresponding vertex of the image using the same transformation, or sequence of transformations. The angle and segment relationships between preimage, image, lines of reflection, center of rotation, and translation vectors can help establish specific transformations that map one point onto another:

- If the segments joining corresponding vertices of the two triangles are congruent and parallel, then one triangle is a translation of the other.
- If a single line is the perpendicular bisector of segments formed by corresponding vertices, then one triangle is a reflection of the other.
- If angles formed by corresponding vertices and a center point are all congruent, and corresponding distances to the center point are equal, then one triangle is a rotation of the other.

For example, given $\overline{ADBE}$, $\overline{AD} \cong \overline{BE}$, $\overline{AD} \cong \overline{CF}$, $\overline{AD} \parallel \overline{CF}$, we can show that $\triangle ABC \cong \triangle DEF$ because one is a translation of the other. Point A translates to point D along segment $\overline{ADBE}$. B translates to E by the same distance along the same segment, and C translates by the same distance along a parallel segment. Each point in $\triangle ABC$ is mapped to a corresponding point in $\triangle DEF$ by the same translation (distance AD along a vector parallel to $\overline{ADBE}$). Therefore, $\triangle DEF$ is a translation of $\triangle ABC$, and the two triangles are congruent because translations are a rigid motion.

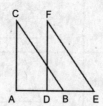

TRIANGLE CONGRUENCE POSTULATES

We know two polygons are congruent if all pairs of corresponding sides and angles are congruent. When proving two triangles congruent we do not need to prove all three pairs of sides and all three pairs of angles congruent. There are five shortcuts commonly used—the SAS, SSS, ASA, AAS, and HL criteria. Each criterion requires only a total of three pairs of parts to be congruent for us to conclude the triangles are congruent.

Criteria	What It Looks Like
SAS—two pairs of sides and the included angles are congruent.	
SSS—three pairs of sides are congruent.	
ASA—two pairs of angles and the included sides are congruent.	
AAS—two pairs of angles and the nonincluded sides are congruent.	
HL—the hypotenuses and one pair of legs are congruent in a pair of right triangles.	

Once a pair of triangles has been proven congruent using one of these postulates, then CPCTC can be used to justify why any of the remaining pairs of corresponding parts are congruent.

You do not need to limit yourself strictly to a rigid motion or a congruence postulate. Sometimes the given information will lead to a combination of these approaches.

STRATEGIES FOR WRITING CONGRUENCE PROOFS

1. Mark congruent parts on the figure using given information and anything else you can conclude from the figure.
2. Look for the following common relationships:
   a. Vertical angles
   b. Shared sides and angles
   c. Linear pairs
   d. Supplementary and complementary angles
   e. Parallel lines and perpendicular lines
   f. Midpoints and bisectors of segments
   g. Angle bisectors
   h. Altitudes, medians, and perpendicular bisectors in triangles
   i. Isosceles triangle theorem, exterior angle theorem, and triangle angle sum theorem
3. Statements should always refer to specific named parts of the figure (such as $\triangle ABC$, $\overline{AB}$, $\angle C$). Reasons may only involve definitions, postulates, and theorems. Never name a specific part of the figure in the reasons column.
4. There must be a line in your proof for each part of the postulate you use. You can label each line that corresponds to a part of the postulate with (A) or (S), or use checkboxes:

S	A	S
✓	✓	✓

OVERLAPPING TRIANGLES

Angle addition or segment addition can be used to show parts are congruent when triangles overlap. For example, if given $\overline{ABCD}$ and $\overline{AB} \cong \overline{CD}$, we can prove $\overline{AC} \cong \overline{BD}$ by stating that $\overline{BC}$ is congruent to itself by the reflexive property, then adding the two congruence statements:

$\overline{BC} \cong \overline{BC}$	reflexive property
$\overline{AC} \cong \overline{BD}$	given
$\overline{AB} + \overline{BC} \cong \overline{BC} + \overline{CD}$	addition property
$\overline{AC} \cong \overline{BD}$	partition property

$$\overset{\displaystyle A \qquad B \qquad C \qquad D}{\bullet\!\!-\!\!-\!\!-\!\!\bullet\!\!-\!\!-\!\!-\!\!\bullet\!\!-\!\!-\!\!-\!\!\bullet}$$

Another strategy is to sketch the two triangles separately. By pulling them apart, you may more easily see the parts required to complete the proof.

Practice Exercises

1 $A'B'C'D'$ is the image of $ABCD$ after a rotation about the origin, and $A''B''C''D''$ is the image of $A'B'C'D'$ after a translation. Which of the following must be true?

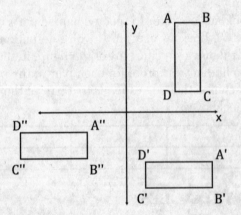

(1) All three figures are rectangles.
(2) $ABCD$ is congruent to $A'B'C'D'$, but not congruent to $A''B''C''D''$.
(3) $ABCD$ is congruent to $A''B''C''D''$, but not congruent to $A'B'C'D'$.
(4) $ABCD$ is congruent to both $A'B'C'D'$ and $A''B''C''D''$.

2 Which triangle congruence criterion can be used to prove the two triangles are congruent?

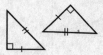

(1) SAS   (3) SSS

(2) ASA   (4) HL

3 Which of the following pieces of information would *not* allow you to conclude that $\triangle ACD \cong \triangle ACB$?

(1) $\overline{AC}$ bisects $\angle BCD$ and $\overline{CD} \cong \overline{CB}$.

(2) $\overline{AC}$ bisects $\angle BCD$ and $\overline{AD} \cong \overline{AB}$.

(3) $\overline{AC}$ bisects $\angle BCD$ and $\angle BAD$.

(4) $\overline{AC}$ bisects $\angle BAD$ and $\overline{AD} \cong \overline{AB}$.

4 Which of the following is sufficient to prove $\triangle SMA \cong \triangle S'M'A'$?

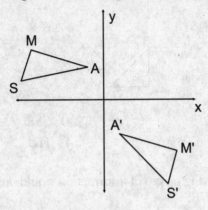

(1)   There is a sequence of rigid motions that maps
      $\angle M$ to $\angle M'$ and $\angle S$ to $\angle S'$.

(2)   There is a sequence of rigid motions that maps
      $\angle M$ to $\angle M'$ and $\overline{MS}$ to $\overline{M'S'}$.

(3)   There is a sequence of rigid motions that maps
      $\angle S$ to $\angle S'$ and $\overline{MS}$ to $\overline{M'S'}$.

(4)   There is a sequence of rigid motions that maps
      $\angle A$ to $\angle A'$ and point A to point A'.

5 Given: $\overline{AB} \parallel \overline{CD}$, $\overline{AB} \cong \overline{CD}$

   Prove: $\triangle ABE \cong \triangle CDE$

6 Given: $\overline{BC}$ bisects $\overline{AD}$ at E, $\angle A \cong \angle D$

   Prove: $\triangle ABE \cong \triangle DCE$

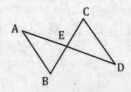

7 *ABCDEF* is a regular hexagon. State a sequence of rigid motions that could be used to justify why △*ABC* is congruent to △*AEF*.

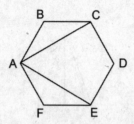

8 In the accompanying figure of △*FGH* and △*GIJ*, there is a sequence of rigid motions that maps *F* to *G*, *G* to *I*, and *H* to *J*. Explain why ∠*H* must be congruent to ∠*J*.

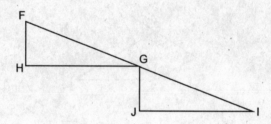

9 Given $\overline{RPB}$ and line *m* is the perpendicular bisector of $\overline{TJ}$ and $\overline{RB}$, explain why △*TRP* ≅ △*JBP* and why $\overline{TR} \cong \overline{JB}$.

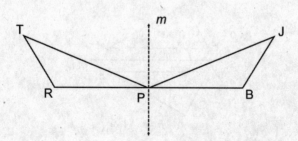

10 Given: $\triangle WXY$ and perpendicular bisector $\overline{YZ}$

Explain why $\triangle WZY \cong \triangle XZY$ in terms of a rigid motion.

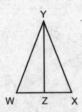

11 Given: $\overline{AB} \cong \overline{AD}$, $\overline{AC} \cong \overline{AE}$, m$\angle BAD = 60°$, and m$\angle CAE = 60°$

Prove: $\triangle BAC \cong \triangle DAE$ in terms of a rigid motion

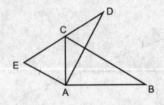

12 Given: line m is the perpendicular bisector of $\overline{FX}$, $\overline{GY}$, and $\overline{HZ}$

Prove: $\triangle XYZ \cong \triangle FHG$

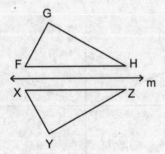

Solutions

1 Both rotations and translations are rigid motions, and any sequence of rigid motions will result in a congruent image. Therefore, $ABCD$ is congruent to both $A'B'C'D'$ and $A''B''C''D''$.

The correct choice is **(4)**.

2 Both triangles are right triangles, a corresponding pair of legs are congruent, and the two hypotenuses are congruent. Therefore, HL applies. SAS is not correct here because the angle does not lie between the two congruent sides.

The correct choice is **(4)**.

3 The shared side $\overline{AC}$ is a pair of congruent sides. Check each choice to find one that does *not* provide enough information to complete one of the congruent postulates.

Choice 1: The angle bisector gives $\angle DCA \cong \angle BCA$, and SAS applies.

Choice 2: The angle bisector gives $\angle DCA \cong \angle BCA$, and we have two sides and an angle, but the angle is not the included angle. None of the congruence postulates apply.

Choice 3: The angle bisector gives $\angle DCA \cong \angle BCA$ and $\angle DAC \cong \angle BAC$. ASA applies.

Choice 4: The angle bisector gives $\angle DAC \cong \angle BAC$, and SAS applies.

The correct choice is **(2)**.

4 Choice 1 maps two consecutive angles onto corresponding angles, as well as M to M' and S to S'. This means $\overline{SM} \cong \overline{S'M'}$ because a sequence of rigid motions maps the endpoints of $\overline{SM}$ onto corresponding endpoints. $\overline{SM}$ is the included side between the congruent angles; therefore, $\triangle SMA \cong \triangle S'M'A'$ by the SAS postulate.

The correct choice is **(1)**.

5

Statement	Reason
1. $\overline{AB} \cong \overline{CD}$, $\overline{AB} \parallel \overline{CD}$	1. Given
2. $\angle A \cong \angle C$ $\angle B \cong \angle D$	2. Alternate interior angles formed by the parallel lines are congruent
3. $\triangle ABE \cong \triangle CDE$	3. ASA

6

Statement	Reason
1. $\overline{BC}$ bisects $\overline{AD}$ at E	1. Given
2. E is the midpoint of $\overline{AD}$	2. A bisector intersects a segment at its midpoint
3. $\overline{AE} \cong \overline{ED}$	3. A midpoint divides a segment into two congruent segments
4. $\angle AEB \cong \angle DEC$	4. Vertical angles are congruent
5. $\angle A \cong \angle D$	5. Given
6. $\triangle ABE \cong \triangle DCE$	6. ASA

7 $\triangle ABC$ is mapped to $\triangle AFE$ by a rotation about point A followed by a reflection over $\overline{AE}$. The angle of rotation is $\angle CAE$. From the figure each of the diagonals $\overline{AC}$, $\overline{AE}$, and $\overline{CE}$ are congruent, so $\triangle ACE$ is equilateral, and its angles measure 60°. The angle of rotation is 60°. The sequence of rigid motions is a 60° rotation about point A followed by a reflection over $\overline{AE}$.

8　The two triangles are congruent because there exists a rigid motion that maps each vertex of $\triangle JGH$ to a corresponding vertex $\triangle GIJ$. Since the triangles are congruent, all pairs of corresponding parts are congruent (CPCTC). $\angle G$ and $\angle H$ are congruent corresponding angles.

9　Since line m is the perpendicular bisector of both $\overline{TJ}$ and $\overline{RB}$, it is a line of reflection that maps T to J and R to B. The same reflection will map P to itself because it lies on the line of reflection. A single reflection maps $\triangle TRP$ onto $\triangle JBP$, and reflections are rigid motions, so the two triangles must be congruent. $\overline{TR} \cong \overline{JB}$ because corresponding parts of congruent triangles are congruent.

10　Point Y is the image of itself after a reflection over $\overline{YZ}$ because Y lies on the line of reflection. Point Z is the image of itself for the same reason. X is the image of point W after a reflection over $\overline{YZ}$ because a line of reflection is the perpendicular bisector of the segment joining the preimage and the image after the reflection. Therefore, the triangles are congruent because one is mapped to the other by a reflection over $\overline{YZ}$, which is a rigid motion.

11　$\overline{AB} \cong \overline{AD}$; therefore, D is the image of B after a $60°$ rotation about A. $\overline{AC} \cong \overline{AE}$; therefore, C is the image of E after a $60°$ rotation about A. Point A will map to itself after the same rotation because it is the center of rotation. Therefore, $\triangle BAC$ is the image of $\triangle DAE$ after a $60°$ rotation about A. $\triangle BAC \cong \triangle DAE$ because rotations are rigid motions.

12　A perpendicular bisector of a segment is also the line of reflection that maps one endpoint to the other. Therefore, after a reflection over line m, X is the image of F, Z is the image of H, and Y is the image of G. $\triangle FHG$ is mapped to $\triangle XZY$ by a reflection, which is a rigid motion; therefore, $\triangle XZY \cong \triangle FHG$.

3.6 COORDINATE GEOMETRY

COORDINATE GEOMETRY FORMULAS

Given two points (x_1, y_1) and (x_2, y_2):

- The distance between the points is $\sqrt{\left(x_2 - x_1\right)^2 + \left(y_2 - y_1\right)^2}$

- The midpoint of the segment joining the points is $\left(\dfrac{x_1 + x_2}{2}, \dfrac{y_1 + y_2}{2}\right)$

- The slope of the segment joining the points is $\dfrac{y_2 - y_1}{x_2 - x_1}$, or $\dfrac{rise}{run}$

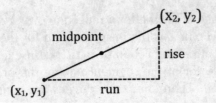

DIVIDING A SEGMENT PROPORTIONALLY

A directed segment is one that has a specified starting point and ending point. We can divide a directed segment into parts that are in any given ratio using the two proportions ratio $= \dfrac{x - x_1}{x_2 - x}$ and ratio $= \dfrac{y - y_1}{y_2 - y}$.

For example, given $J(1, -2)$ and $K(11, 3)$, find the coordinate (x, y) of point L that divides $\overline{JK}$ in a $2 : 3$ ratio.

$$J(1, -2) \qquad L(x, y) \qquad\qquad K(11, 3)$$

- Use the coordinates of J for point 1 and the coordinates of K for point 2.

$$x_1 = 1 \qquad x_2 = 11 \qquad y_1 = -2 \text{ and } y_2 = 3$$

- Apply the formula for the x-coordinate.

$$\text{ratio} = \frac{x - x_1}{x_2 - x}$$

$$\frac{2}{3} = \frac{x - 1}{11 - x}$$
$$2(11 - x) = 3(x - 1)$$
$$22 - 2x = 3x - 3$$
$$25 = 5x$$
$$x = 5$$

- Repeat the process for the y-coordinate.

$$\text{ratio} = \frac{y - y_1}{y_2 - y}$$

$$\frac{2}{3} = \frac{y - (-2)}{3 - y}$$
$$3y + 6 = 6 - 2y$$
$$5y = 0$$
$$y = 0$$

Point $L(5, 0)$ divides $\overline{JK}$ in a $2 : 3$ ratio.

AREA AND PERIMETER

The lengths found with the distance formula can be used to calculate the perimeter and area of figures. If the figure is irregular, three strategies can be used:

- Divide the figure into shapes whose areas can be calculated easily (squares, rectangles, triangles, trapezoids, and circles).
- Sketch a bounding rectangle around the figure. Calculate the area of the rectangle, and then subtract the area of all the triangles that fall outside the figure but with the rectangle.
- If the figure has curves, estimate the area by modeling the curved sides with straight segments. The more segments that are used to model a curve, the more accurate the result will be.

Example

Find the area of polygon $ABCDE$, with vertices $A(-3, -5)$, $B(2, -5)$, $C(5, 3)$, $D(-2, 5)$, and $E(-5, -1)$.

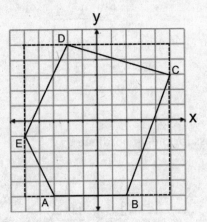

Solution:

Sketch the bounding rectangle in around $ABCDE$.

The length is 10 and the width is 10, giving an area of 100. The triangles have areas:

$$\text{upper left triangle} = \frac{1}{2}(3 \cdot 6) = 9$$

$$\text{upper right triangle} = \frac{1}{2}(7 \cdot 2) = 7$$

$$\text{lower left triangle} = \frac{1}{2}(4 \cdot 2) = 4$$

$$\text{lower right triangle} = \frac{1}{2}(3 \cdot 8) = 12$$

The area of $ABCDE = 100 - 9 - 7 - 4 - 12 = 68$.

COLLINEARITY

Three points are *collinear* if the slopes between any two pairs are equal. For example, points A, B, and C are collinear if the slope of $\overline{AB}$ equals the slope of $\overline{BC}$.

EQUATIONS OF LINES

- The slopes of parallel lines are equal.
- The slopes of perpendicular lines are negative reciprocals. (If the slope of line m is $\frac{2}{3}$, then the slope of any line perpendicular to m is $-\frac{3}{2}$.)
- The equation of a line in slope-intercept form is $y = mx + b$ where m is the slope and b is the y-intercept. To graph the line, plot a point on the y-axis at the y-intercept. From that point, plot additional points using the rise and run from the slope.
- The equation of a line in point-slope form is $y - y_1 = m(x - x_1)$ where m is the slope and (x_1, y_1) are the coordinates of any point on the line. To graph the line, plot the first point at (x_1, y_1). From that point, plot additional points using the rise and run from the slope.

Strategy for writing the equation of a line in slope-intercept form:

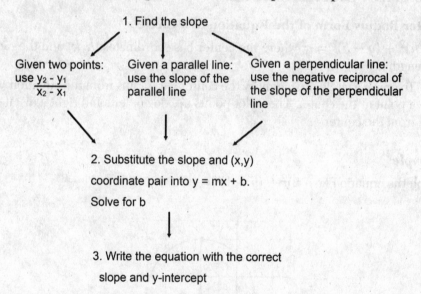

1. Find the slope

Given two points: use $\frac{y_2 - y_1}{x_2 - x_1}$

Given a parallel line: use the slope of the parallel line

Given a perpendicular line: use the negative reciprocal of the slope of the perpendicular line

2. Substitute the slope and (x,y) coordinate pair into y = mx + b. Solve for b

3. Write the equation with the correct slope and y-intercept

TRANSFORMATIONS AND LINES

Translations and Dilations

Translations and dilations preserve slope, so the slope of the image will be the same as the slope of the preimage.

To translate or dilate a line given its equation,

1. Choose any point on the line (the y-intercept is often an easy choice).

2. Apply the translation or dilation to that point.

3. Find the equation of the line that has the same slope as the original line and passes through the transformed point.

Rotations

Rotations of 90° will result in a line perpendicular to the original, so the slope will be the negative reciprocal. To write the equation of a line after a 90° rotation, use the same procedure for translations and dilations, except use the negative reciprocal of the slope.

EQUATION OF THE CIRCLE

Center Radius Form of the Equation of a Circle

$(x - h)^2 + (y - k)^2 = r^2$ where the center has coordinates (h, k) and the radius has length r.

- To graph a circle, first identify the center and radius from the equation. Plot a point at the center. Then plot points up, down, left, and right a distance r from the center.

Example

Graph the equation $(x - 2)^2 + (y + 1)^2 = 9$.

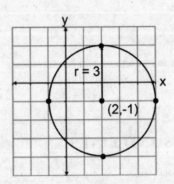

The center is located at $(2, -1)$, and $r^2 = 9$, so $r = 3$. We plot the center point at $(2, -1)$; then plot points up, down, right, and left 3 units from the center. Use these four points as a guide to complete the circle.

General Form of the Equation of a Circle

$$x^2 + y^2 + Cx + Dy + E = 0$$

To find the coordinates of the center and the radius from the general form of the equation, you will need to convert it to the center-radius form using the following procedure:

1. Group the x-terms and y-terms on one side of the equation, and the constant on the other side of the equation.

2. Complete the square with the x-terms, and then complete the square with the y-terms.

Example

Find the coordinates of the center and the length of the radius of a circle whose equation is $x^2 + 4x + y^2 - 6y + 7 = 0$.

Solution:

Bring the constant term to the right.

$$x^2 + 4x + y^2 - 6y = -7$$

The coefficient of x is 4, so a constant term of $\left(\frac{4}{2}\right)^2$, or 4, is needed to complete the square with the x-terms. The coefficient of y is -6, so a constant term of $\left(\frac{-6}{2}\right)^2$, or 9, is needed to complete the square with the y-terms.

$$x^2 + 4x + 4 + y^2 - 6y + 9 = -7 + 4 + 9$$
$$(x + 2)^2 + (y - 3)^2 = 6$$

The center has coordinates $(-2, 3)$ and the radius has a length of $\sqrt{6}$.

Practice Exercises

1 Points $A(2, -1)$ and $B(8, -3)$ lie on line m. After a rotation of 90° about the origin, the images of A and B are A' and B'. If A' and B' lie on line n, what is the equation of line n?

(1) $y = -3x + 2$ (3) $y = 3x - 1$

(2) $y = -\dfrac{1}{3}x - \dfrac{1}{3}$ (4) $y = \dfrac{1}{3}x + 6$

2 What is the equation of the line $6x + 2y = 12$ after a dilation by a scale factor of 5?

(1) $y = -3x + 30$ (3) $y = -15x + 30$

(2) $y = -3x + 6$ (4) $y = -15x + 6$

3 Which of the following lines is perpendicular to the line $x + 4y = 8$?

(1) $y = -\dfrac{1}{4}x + 2$ (3) $y = \dfrac{1}{4}x + 2$

(2) $y = 4x + 3$ (4) $y = -4x + 3$

4 Which of the following is the equation of a line parallel to $2x + 3y + 6 = 0$ and passes through the point $(6, 1)$?

(1) $y = -\dfrac{2}{3}x + 5$ (3) $y = -2x + 13$

(2) $y = 2x - 11$ (4) $y = -\dfrac{2}{3}x + 3$

5 What are the coordinates of the midpoint of a segment whose endpoints have coordinates $(3, 1)$ and $(15, -7)$?

   (1) $(27, -15)$ (3) $(6, -4)$

   (2) $(-6, 4)$ (4) $(9, -3)$

6 The diameter of a circle has endpoints with coordinates $(4, -1)$ and $(8, 3)$. Which of the following is an equation of the circle?

   (1) $(x - 2)^2 + (y - 1)^2 = 8$ (3) $(x - 6)^2 + (y - 1)^2 = 8$

   (2) $(x - 2)^2 + (y - 1)^2 = 32$ (4) $(x - 6)^2 + (y - 1)^2 = 32$

7 Are the segments $\overline{AB}$ and $\overline{TU}$ congruent, given coordinates $A(1, 4)$, $B(-3, 6)$, $T(2, 5)$, and $U(4, 1)$? Justify your answer.

8 Find the coordinates of the point W that divides directed segment $\overline{UV}$ in a $1 : 5$ ratio, given coordinates $U(-3, 7)$ and $V(9, 1)$.

9 Point A has coordinates $(-2, 7)$ and point B has coordinates $(6, 3)$. Line m has the property that every point on the line is equidistant from points A and B. Find the equation of line m.

10 A circle is described by the equation $x^2 + 6x + y^2 - 12y + 25 = 0$. Find the radius of the circle and the coordinates of its center.

11 Circle P has center $P(4, -5)$ and a radius with length $\sqrt{65}$. Does the point $A(8, 2)$ lie on circle P? Justify your answer.

12 Parallelogram $ABCD$ has coordinates $A(2, -1)$, $B(5, 1)$, $C(a, b)$, and $D(3, 4)$. Write the equation of the line that contains side $\overline{CD}$.

13 Estimate the area of the ellipse shown in the accompanying figure by modeling it as the sum of 5 rectangles.

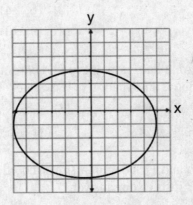

14 Ben wants to estimate the area of the curved region shown. He would like to do so by calculating the average of the areas of two different circles.

a) Graph the two circles that Ben could use.

b) Using the two circles from part (a), estimate the area of the curved region. Each grid unit represents 1 centimeter. Round to the nearest 1 cm^2.

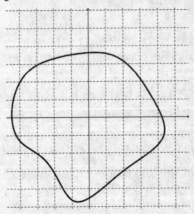

15 $\triangle SKY$ has coordinates $S(0, 2)$, $K(6, 0)$, $Y(4, 4)$.

a) Find the equation of the perpendicular bisector to side $\overline{SK}$.

b) Show that the perpendicular bisector is also an altitude.

Solutions

1 Applying a 90° rotation, A maps to $A'(1, 2)$ and B maps to $B'(3, 8)$.

$$\text{Slope } \overleftrightarrow{A'B'} = \frac{y_2 - y_1}{x_2 - x_1}$$

$$= \frac{8 - 2}{3 - 1}$$

$$= 3$$

Write the equation of the line using the point-slope form and the coordinates of A' for x_1 and y_1.

$$y - y_1 = m\left(x - x_1\right)$$
$$y - 2 = 3(x - 1)$$
$$y - 2 = 3x - 3$$
$$y = 3x - 1$$

The correct choice is **(3)**.

2 A dilation with a positive scale factor will preserve slope, but the distance of each point from the center will be multiplied by the scale factor. The strategy is to rewrite the equation in $y = mx + b$ form. The dilated line will have the same slope, but the y-intercept will be multiplied by the scale factor.

$$6x + 2y = 12$$
$$2y = -6x + 12$$
$$y = 3x + 6$$

The y-intercept of the original line is 6, and the y-intercept of the dilated line is $6 \cdot 5$, or 30.

$$y = -3x + 30$$

The correct choice is **(1)**.

3 Perpendicular lines have negative reciprocal slopes. Rewrite the equation in slope-intercept form to identify the slope.

$$x + 4y = 8$$
$$4y = -x + 8$$
$$y = -\frac{1}{4}x + 2$$

The negative reciprocal of $-\frac{1}{4}$ is 4, so the perpendicular line must have a slope of 4. The y-intercept can be any value since a particular perpendicular line was not specified. The only line with a slope of 4 is $y = 4x + 3$.

The correct choice is (2).

4 Parallel lines have equal slopes, so the first step is to rewrite the equation in slope-intercept form to help identify the slope.

$$2x + 3y + 6 = 0$$
$$3y = -2x - 6$$
$$y = -\frac{2}{3}x - 2$$

The slope is $-\frac{2}{3}$.

Now substitute $x = 6$, $y = 1$, and the slope into $y = mx + b$ and solve for the new y-intercept.

$$y = mx + b$$
$$y = -\frac{2}{3}x + b$$
$$1 = -\frac{2}{3}(6) + b$$
$$1 = -4 + b$$
$$b = 5$$

The equation is $y = -\frac{2}{3}x + 5$.

The correct choice is (1).

5 Apply the midpoint formula:

$$x_{MP} = \frac{1}{2}(x_1 + x_2) \quad y_{MP} = \frac{1}{2}(y_1 + y_2)$$

$$x_{MP} = \frac{1}{2}(3 + 15) \quad y_{MP} = \frac{1}{2}(1 + (-7))$$

$$x_{MP} = \frac{1}{2}(18) \quad y_{MP} = \frac{1}{2}(-6)$$

$$x_{MP} = 9 \quad y_{MP} = -3$$

The correct choice is **(4)**.

6 The center and radius of the circle are needed to write its formula. The midpoint of the diameter gives the center:

$$x_{MP} = \frac{1}{2}(x_1 + x_2) \quad y_{MP} = \frac{1}{2}(y_1 + y_2)$$

$$x_{MP} = \frac{1}{2}(4 + 8) \quad y_{MP} = \frac{1}{2}(-1 + 3)$$

$$x_{MP} = \frac{1}{2}(12) \quad y_{MP} = \frac{1}{2}(2)$$

$$x_{MP} = 6 \quad y_{MP} = 1$$

The radius is the distance from the center point to either endpoint of the diameter. Apply the distance formula with points (6, 1) and (8, 3).

$$\text{distance} = \sqrt{(x_1 - x_2)^2 + (y_1 - y_2)^2}$$
$$= \sqrt{(8 - 6)^2 + (3 - 1)^2}$$
$$= \sqrt{(2)^2 + (2)^2}$$
$$= \sqrt{8}$$

The radius of the circle is $\sqrt{8}$, and its center has coordinates $(6, 1)$. Substitute these values for r, h, and k in the equation of a circle:

$$(x - h)^2 + (y - k)^2 = R^2$$
$$(x - 6)^2 + (y - 1)^2 = \sqrt{8}^2$$
$$(x - 6)^2 + (y - 1)^2 = 8$$

The correct choice is **(3)**.

7 Two segments are congruent if their lengths are equal, so apply the distance formula to determine the length of each segment.

$$d = \sqrt{(x_1 - x_2)^2 + (y_1 - y_2)^2}$$

$$AB = \sqrt{(-3 - 1)^2 + (6 - 4)^2} \qquad TU = \sqrt{(4 - 2)^2 + (1 - 5)^2}$$
$$= \sqrt{(-4)^2 + (2)^2} \qquad\qquad = \sqrt{(2)^2 + (-4)^2}$$
$$= \sqrt{16 + 4} \qquad\qquad\qquad = \sqrt{4 + 16}$$
$$= \sqrt{20} \qquad\qquad\qquad\quad = \sqrt{20}$$

$AB = TU$; therefore, the 2 segments are congruent.

8

$$\frac{UW}{WV} = \frac{1}{5} = \frac{x - (-3)}{9 - x}$$
$$9 - x = 5(x + 3)$$
$$9 - x = 5x + 15$$
$$-6 = 6x$$
$$x = -1$$

Repeating for the y-coordinate:

$$\frac{UW}{WV} = \frac{1}{5} = \frac{y - 7}{1 - y}$$
$$1 - y = 5(y - 7)$$

The coordinates of W are $(-1, 6)$.

9 Line m is the line of reflection that maps A to B, so it must be the perpendicular bisector of $\overline{AB}$. To find the perpendicular bisector, calculate the midpoint and slope of $\overline{AB}$. Then write the equation of the line with the negative reciprocal slope that passes through the midpoint.

$$x_{MP} = \frac{1}{2}(x_1 + x_2), \quad y_{MP} = \frac{1}{2}(y_1 + y_2)$$

$$x_{MP} = \frac{1}{2}(-2 + 6), \quad y_{MP} = \frac{1}{2}(7 + 3)$$

$$x_{MP} = 2, \quad y_{MP} = 5$$

$$\text{slope} = \frac{y_2 - y_1}{x_2 - x_1}$$

$$\text{slope}_{AB} = \frac{3 - 7}{6 - (-2)}$$

$$= -\frac{1}{2}$$

$\text{slope}_{\text{line } m} = 2$ $\perp$ lines have negative reciprocal slopes

$y - y_1 = m(x - x_1)$ point-slope equation of a line

$y - 5 = 2(x - 2)$ or $y = 2x + 1$ substitute the coordinates of the midpoint for x_1 and y_1, and 2 for m

10 Apply the completing the square procedure to rewrite the circle in $(x - h)^2 + (y - k)^2 = R^2$ form. Rewrite the equation with the variables on the left and constant on the right.

$$x^2 + 6x + y^2 - 12y + 25 = 0$$

$$x^2 + 6x + y^2 - 12y = -25$$

The constant needed to complete the square is $\left(\frac{1}{2}b\right)^2$, where b is the coefficient of the linear x- and y-terms. For the x-terms, the necessary constant is $\left(\frac{1}{2}(6)\right)^2$, or 9. For the y-terms $\left(\frac{1}{2}(-12)\right)^2$, or 36, is needed. Add the required constants to each side of the equation.

$$x^2 + 6x + 9 + y^2 - 12y + 36 = -25 + 9 + 36$$

$$(x + 3)^2 + (y - 6)^2 = 20 \quad \text{factor the } x\text{-terms and the } y\text{-terms}$$

The center has coordinates $(-3, 6)$ and the radius is $\sqrt{20}$.

11 A point that lies on a curve will satisfy the equation of the curve, so the strategy is to write the equation of the circle. Then substitute the coordinates of point A for x and y in the equation and check if the equation is balanced.

$$(x - h)^2 + (y - k)^2 = R^2 \quad \text{equation of a circle}$$

$$(x - 4)^2 + (y + 5)^2 = \sqrt{65}^2 \quad \text{substitute } h = 4, k = -5, \text{ and } R = \sqrt{65}$$

$$(8 - 4)^2 + (2 + 5)^2 = \sqrt{65}^2 \quad \text{substitute coordinates of } A \text{ for } x \text{ and } y$$

$$4^2 + 7^2 = \sqrt{65}^2$$

$$16 + 49 = 65$$

$$65 = 65$$

The equation is balanced, so the point $A(8, 2)$ does lie on circle P.

12 Opposite sides of a parallelogram are parallel, so side $\overline{CD}$ is parallel to side $\overline{AB}$ and must have the same slope.

Start by finding the slope of $\overline{AB}$.

$$\text{slope } \overline{AB} = \frac{y_2 - y_1}{x_2 - x_1}$$

$$= \frac{1 - (-1)}{5 - 2}$$

$$= \frac{2}{3}$$

The desired line also passes through point $D(3, 4)$; $x = 3$ and $y = 4$ must satisfy the equation of $\overline{CD}$. Apply the point-slope equation of a line with slope $= \frac{2}{3}$, which represents the slope of $\overline{AB}$ using $x_1 = 3$ and $y_1 = 4$:

$$y - y_1 = m\left(x - x_1\right)$$
$$y - 4 = \frac{2}{3}(x - 3)$$

or, in slope-intercept form, $y = \frac{2}{3}x + 2$.

13 We can estimate the area by filling in the ellipse with 5 rectangles. Triangles could have been used to get a more accurate estimate of the area.

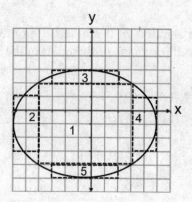

The areas of the rectangles are

    rectangle 1 area $= 7 \cdot 6 = 42$

    rectangle 2 area $= 4 \cdot 2 = 8$

    rectangle 3 area $= 5 \cdot 1 = 5$

    rectangle 4 area $= 4 \cdot 2 = 8$

    rectangle 5 area $= 5 \cdot 1 = 5$

The approximate area of the ellipse is $42 + 8 + 5 + 8 + 5 = 68$ square units.

14 The two circles are the largest that will fit inside the region, and the smallest that will fit outside the region.

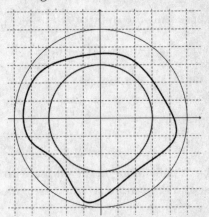

The inner circle has a radius of 3 cm and the outer circle has a radius of 5 cm. The estimate is the average of the areas of the two circles.

$$A_{\text{estimate}} = \frac{1}{2}\left(\pi R^2_{\text{inner}} + \pi R^2_{\text{outer}}\right)$$

$$= \frac{1}{2}\left(\pi \cdot 3^2 + \pi \cdot 5^2\right)$$

$$= \frac{1}{2}(106.8141)$$

$$\approx 53.4 \text{ cm}^2$$

15 a) To find the equation of a perpendicular bisector, the slope and midpoint of $\overline{SK}$ are needed.

$$\text{slope } \overline{SK} = \frac{y_2 - y_1}{x_2 - x_1}$$

$$= \frac{0 - 2}{6 - 0}$$

$$= -\frac{1}{3}$$

midpoint $\overline{SK}$ is $\frac{x_1 + x_2}{2}, \frac{y_1 + y_2}{2}$

$$\left(\frac{0 + 6}{2}, \frac{2 + 0}{2}\right)$$

$$(3, 1)$$

Perpendicular lines have negative reciprocal slopes, so the slope of the perpendicular bisector is 3, and the line must pass through the midpoint $(3, 1)$. Substitute these values in the point-slope equation of a line:

$$y - y_1 = m(x - x_1)$$
$$y - 1 = 3(x - 3)$$

or $y = 3x - 8$ in slope-intercept form.

b) An altitude will pass through the opposite vertex, so check if the coordinates of point $Y(4, 4)$ satisfy the equation of the perpendicular bisector.

$$y - 1 = 3(x - 3)$$
$$4 - 1 = 3(4 - 3)$$
$$3 = 3(1)$$
$$3 = 3$$

The equation is balanced, so the perpendicular bisector passes through the opposite vertex and is also an altitude.

3.7 SIMILAR FIGURES

PROPERTIES AND DEFINITION OF SIMILAR FIGURES

Two figures are similar if there exists a sequence of similarity transformations that maps one figure onto another. Remember, the sequence of similarity transformations often includes a dilation.

Similar figures have the following properties:

- Same shape but possibly different sizes
- All corresponding lengths proportional
- All corresponding angles congruent

The ratio of proportionality is called the *scale factor*. A similarity statement is written using the symbol ~. In the accompanying figure, $\triangle ABC \sim \triangle DEF$, from the similarity statement, we have the following relationships:

$$\angle A \cong \angle D \quad \angle B \cong \angle E \quad \angle C \cong \angle F$$

$$\frac{AB}{DE} = \frac{BC}{EF} = \frac{AC}{DF} = \frac{1}{3}$$

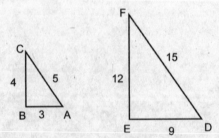

PROVING TRIANGLES SIMILAR USING THE SIMILARITY POSTULATES

- Two triangles can be proven similar using the following postulates:
  AA (angle–angle)
  SAS (side–angle–side)
  SSS (side–side–side)

Unlike congruence, pairs of corresponding sides must be in the same ratio.

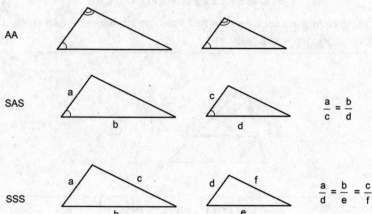

AA

SAS $\dfrac{a}{c} = \dfrac{b}{d}$

SSS $\dfrac{a}{d} = \dfrac{b}{e} = \dfrac{c}{f}$

PERIMETER, AREA, AND VOLUME IN SIMILAR FIGURES

Perimeter, area, and volume in similar figures and solids are related to the ratio of corresponding sides.

$\dfrac{\text{Perimeter}_A}{\text{Perimeter}_B}$	$\dfrac{\text{Area}_A}{\text{Area}_B}$	$\dfrac{\text{Volume}_A}{\text{Volume}_B}$
scale factor	(scale factor)^2	(scale factor)^3

SIMILARITY RELATIONSHIPS IN TRIANGLES

SEGMENT PARALLEL TO A SIDE THEOREM
- A segment parallel to a side of a triangle forms a triangle similar to the original triangle.
- If a segment intersects two sides of a triangle such that a triangle similar to the original is formed, the segment is parallel to the third side of the original triangle.

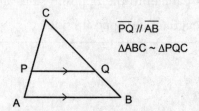

$\overline{PQ} \,/\!/\, \overline{AB}$

$\triangle ABC \sim \triangle PQC$

SIDE SPLITTER THEOREM

A segment parallel to a side in a triangle divides the two sides it intersects proportionally.

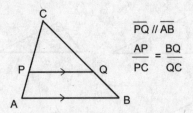

$$\overline{PQ} \, / \! / \, \overline{AB}$$

$$\frac{AP}{PC} = \frac{BQ}{QC}$$

CENTROID THEOREM

The centroid of a triangle divides each median in a 1 : 2 ratio, with the longer segment having a vertex as one of its endpoints.

In the accompanying figure, D is the midpoint of $\overline{AB}$, and E is the midpoint of $\overline{BC}$, making $\overline{AE}$ and $\overline{CD}$ medians. Their point of intersection, G, is the centroid and divides the medians in a 1 : 2 ratio.

$$\frac{EG}{AG} = \frac{1}{2} \text{ and } \frac{DG}{CG} = \frac{1}{2}$$

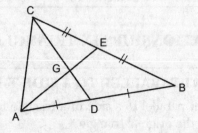

MIDSEGMENT THEOREM

A segment joining the midpoints of two sides of a triangle (a midsegment) is parallel to the opposite side, and its length is equal to $\frac{1}{2}$ the length of the opposite side.

In the accompanying figure, E is the midpoint of $\overline{AB}$, and F is the midpoint of $\overline{AC}$. $\overline{EF} \parallel \overline{BC}$ and $EF = \frac{1}{2}BC$.

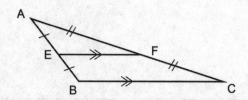

ALTITUDE TO THE HYPOTENUSE OF A RIGHT TRIANGLE THEOREM

The altitude to the hypotenuse of a right triangle forms two triangles that are similar to the original triangle.

In the accompanying figure, $\overline{CD}$ is an altitude to hypotenuse $\overline{AB}$, $\triangle BDC \sim \triangle CDA \sim \triangle BCA$.

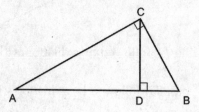

Two useful strategies for dealing with these overlapping similar triangles is to sketch the triangles separately or to make a table with sides classified as the "leg 1," "leg 2," and "hypotenuse." The altitude will be leg 1 of one of the interior triangles, and leg 2 of the other.

	Leg 1	Leg 2	Hypotenuse
Small △	$\overline{BD}$	$\overline{CD}$	$\overline{BC}$
Medium △	$\overline{CD}$	$\overline{AD}$	$\overline{AC}$
Large △	$\overline{BC}$	$\overline{AC}$	$\overline{AB}$

Practice Exercises

1 In the accompanying figure, $\triangle PQR \sim \triangle TSR$ and points P, R, and T are collinear. Which of the following transformations will map $\triangle TSR$ to $\triangle PQR$?

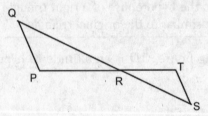

(1) reflection over line $\overline{TP}$ followed by a dilation with a scale factor of $\dfrac{QR}{TR}$ centered at Q

(2) translation by vector $\overline{TR}$ followed by a dilation with a scale factor of $\dfrac{PQ}{ST}$ centered at R

(3) rotation of $180°$ about point R followed by a dilation with a scale factor of $\dfrac{PQ}{ST}$ centered at R

(4) rotation of $180°$ about point R followed by a dilation with a scale factor of $\dfrac{QR}{TR}$ centered at R

2 $ABCDE$ and $VWXYZ$ are similar pentagons. If $BC = 4$ and $WX = 6$, what is the ratio of the area of $ABCDE$ to the area of $VWXYZ$?

(1) $\dfrac{4}{9}$

(2) $\dfrac{2}{3}$

(3) $\dfrac{2}{5}$

(4) $\sqrt{\dfrac{2}{3}}$

3 If $\triangle XYZ \sim \triangle XVW$, which of the following transformations will map $\triangle XVW$ to $\triangle XYZ$?

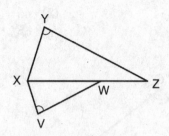

(1) reflect $\triangle XVW$ over $\overline{XW}$ and then dilate it with a center at X and scale factor of $\dfrac{XW}{WZ}$

(2) reflect $\triangle XVW$ over $\overline{XW}$ and then dilate it with a center at X and scale factor of $\dfrac{XZ}{XW}$

(3) rotate $\triangle XVW$ 180° about point X and then dilate it with a center at X and scale factor of $\dfrac{XW}{WZ}$

(4) rotate $\triangle XVW$ 180° about point X and then dilate it with a center at X and scale factor of $\dfrac{XZ}{XW}$

4 Which of the following triangles can be proven to be similar?

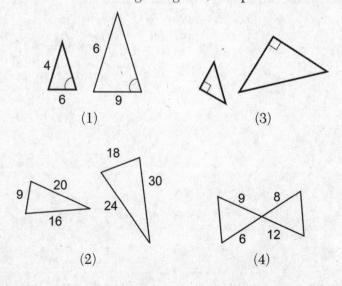

5 Given triangles FOG and BAT with $\angle G \cong \angle T$, $\angle O \cong \angle B$, $FG = 5x$, $FO = 42$, $AB = 14$, and $AT = x + 12$, find the length FG.

(1) 18   (3) 45

(2) 30   (4) 90

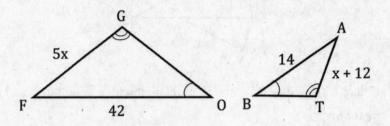

6 $\overline{UV}$ is parallel to side $\overline{ST}$ of $\triangle RST$. If $RU = x - 6$, $US = x$, $RV = x$, and $VT = 2x - 9$, what is the value of x?

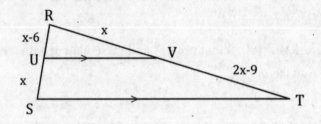

(1) 3   (3) 18

(2) 11   (4) 39

7 In the figure below, D lies on $\overline{AB}$ and E lies on $\overline{AC}$. If $AD = 3$, $BD = x + 3$, $DE = 5$, and $BC = 2x + 2$, find the value of x that would let you conclude that $\overline{DE} \parallel \overline{BC}$.

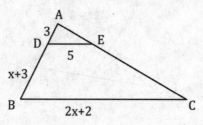

  (1) 2   (3) 18

  (2) 9   (4) 24

8 Mollie wants to rearrange the furniture in her living room. She measures the rectangular room to be 16 ft by 12 ft and makes a scaled drawing of the room that measures 10 in by 7.5 in. She wants to make scaled paper cutouts to represent her furniture so she can determine her favorite arrangement without having to move the furniture. If her sofa is rectangular and measures 6 ft by 2 ft, what should be the dimensions of the cutout?

  (1) $1\frac{1}{3}$ in $\times$ 1 in   (3) $3\frac{3}{4}$ in $\times 1\frac{1}{4}$ in

  (2) $2\frac{3}{4}$ in $\times$ 1 in   (4) 9.6 in $\times$ 3.2 in

9 Given: $\overline{LSI}$ and $\overline{FSP}$ intersecting at S, $\overline{FL} \parallel \overline{IP}$, $IP = 4$ and $FL = 6$

  a) Explain why the two triangles formed are similar.

  b) Find the ratio $\dfrac{\overline{PS}}{\overline{FS}}$.

  c) State a specific similarity transformation that will map the smaller triangle onto the larger.

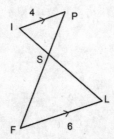

10 In $\triangle VTP$, F is the midpoint of $\overline{VP}$, G is the midpoint of $\overline{PT}$, and $\overline{VG}$ and $\overline{FT}$ intersect at K. If $VK = 5x + 5$ and $KG = 2x + 8$, find the length VG.

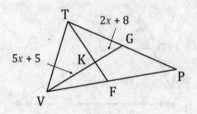

11 The midpoints of $\overline{PG}$, $\overline{PI}$, and $\overline{GI}$ are C, O, and W, respectively. If $OW = 7$, $CW = 5$, and $OC = 6$, find the perimeter of $\triangle PIG$.

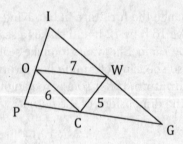

12 In the figure below, $\angle ABC$ and $\angle BDC$ are right angles, $CB = 6$, and $AB = 12$. Find the length of $\overline{BD}$. Express your answer in simplest radical form.

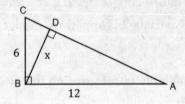

13 Rectangle *AEFG* is the image of rectangle *ABCD* after the transformation clockwise $R_{90°,A} \circ D_{\frac{2}{3},A}$. If $GA = 8$ and $DE = 12$, what is the perimeter of *ABCD*?

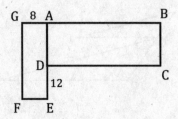

14 *ABCDEF* and *ARSTUV* are both regular hexagons and R is the midpoint of $\overline{AB}$. Name a sequence of transformations that will map *ARSTUV* to *ABCDEF*.

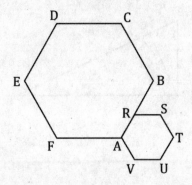

15 Given: $\overline{RUT}$, $\overline{SVT}$, and $\overline{UV} \parallel \overline{RS}$

Prove: $\dfrac{UR}{TU} = \dfrac{VS}{TV}$

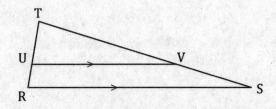

16 Given: $\overline{ABD}$, $\angle C \cong \angle E$, $\angle DEB \cong \angle EBC$

Prove: $\triangle ABC \sim \triangle BDE$

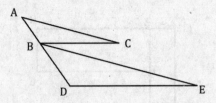

17 Prove the Pythagorean theorem.

Given: right triangle $\triangle BCA$ with a right angle at C, altitude $\overline{CD}$ drawn to hypotenuse $\overline{AB}$

Prove: $(AC)^2 + (BC)^2 = (AB)^2$

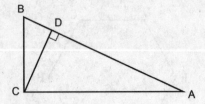

Solutions

1 We want to map the smaller triangle to the larger, so the scale factor will be greater than 1. The correct scale factor among the choices is $\frac{PQ}{ST}$. A rotation of $180°$ will rotate $\triangle TSR$ to $\triangle PQR$. The correct sequence is: a rotation of $180°$ centered at R, followed by a dilation by a scale factor of $\frac{PQ}{ST}$ centered at R. Note that the order of the sequence does not matter in this situation.

The correct choice is **(3)**.

2 The ratio of the areas of similar figures is equal to the ratio of lengths of corresponding sides squared. $\overline{BC}$ and $\overline{WX}$ are corresponding sides, and are in a ratio of $\frac{4}{6}$, or $\frac{2}{3}$.

$$\text{ratio of areas} = \text{ratio of sides}^2$$
$$= \left(\frac{2}{3}\right)^2$$
$$= \frac{4}{9}$$

The correct choice is **(1)**.

3 A reflection over $\overline{XW}$ will map $\overline{XV}$ onto $\overline{XZ}$. Then a dilation centered about X with a scale factor of $\frac{XZ}{XW}$ will match the sizes.

The correct choice is **(2)**.

4 The triangle similarity postulates are **AA**, **SAS**, and **SSS**, where each S represents the *ratio* of a pair of corresponding sides. Examine each of the choices.

Choice 1: The two pairs of corresponding sides are in a 2 : 3 ratio, but the congruent angle is not the included angle.

Choice 2: Corresponding sides are not in the same ratio, $\frac{9}{18} \neq \frac{16}{24}$.

Choice 3: Lengths of the sides are not given.

Choice 4: 2 pairs of corresponding sides are in the same ratio, $\frac{6}{8} = \frac{9}{12}$, and the congruent vertical angles are the included angles. The triangles are similar by SAS.

The correct choice is **(4)**.

5 The two triangles are similar by AA, and corresponding sides of similar triangles are in the same ratio. The similarity statement can be written as $\triangle FOG \sim \triangle ABT$, so the following proportion can be written:

$$\frac{FG}{AT} = \frac{FO}{AB}$$
$$\frac{5x}{x+12} = \frac{42}{14}$$
$$14(5x) = 42(x+12) \quad \text{cross products are equal}$$
$$70x = 42x + 504$$
$$28x = 504$$
$$x = 18 \quad\quad\quad \text{substitute } x = 18$$
$$FG = 5(18)$$
$$= 90$$

The correct choice is **(4)**.

6 When a segment is drawn parallel to a side of a triangle, the two triangles are similar, and the side splitter theorem states that the sides are divided proportionally.

$$\frac{RU}{US} = \frac{RV}{VT} \quad\quad \text{side splitter theorem}$$
$$\frac{x-6}{x} = \frac{x}{2x-9}$$

$$(2x - 9)(x - 6) = x^2 \quad \text{cross products are equal}$$
$$2x^2 - 21x + 54 = x^2$$
$$x^2 - 21x + 54 = 0 \quad \text{solve by moving all terms to one side and factoring}$$
$$(x - 18)(x - 3) = 0 \quad \text{factor the trinomial}$$
$$x - 18 = 0 \quad x - 3 = 0 \quad \text{zero product theorem}$$
$$x = 18 \quad\quad x = 3$$

The two solutions of the quadratic are *possible* solutions to the problem. Eliminate any results that lead to negative lengths or any other impossible situations. The solution $x = 3$ would result in a negative length for RU and VT, so we eliminate this result, and the value of x is 18.

The correct choice is **(3)**.

7 From the original figure, $\overline{DE} \parallel \overline{BC}$ when $\angle CBA \cong \angle EDA$ (corresponding angles are congruent). The triangles $\triangle ADE$ and $\triangle ABC$ are similar by AA, and corresponding sides will be proportional. Sketch the two triangles separately to help identify the corresponding sides. Note that $AB = AD + DB$.

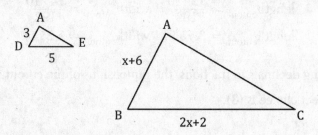

$$\frac{AD}{AB} = \frac{DE}{BC} \quad\quad \text{corresponding sides are proportional}$$
$$\frac{3}{x + 6} = \frac{5}{2x + 2}$$
$$3(2x + 2) = 5(x + 6) \quad \text{cross products are equal}$$
$$6x + 6 = 5x + 30$$
$$x = 24$$

The correct choice is **(4)**.

8 First convert feet to inches so all units are consistent.

$$\text{Length of room} \quad 16 \text{ ft} \times \frac{12 \text{ in}}{\text{ft}} = 192 \text{ in}$$

$$\text{Length of sofa} \quad 6 \text{ ft} \times \frac{12 \text{ in}}{\text{ft}} = 72 \text{ in}$$

$$\text{Width of sofa} \quad 2 \text{ ft} \times \frac{12 \text{ in}}{\text{ft}} = 24 \text{ in}$$

The scaled drawing is similar to the actual room, so the ratio of corresponding sides is

$$\frac{\text{length of drawing}}{\text{length of room}} = \frac{10}{192}$$

Apply this scale factor to the length and width of the sofa to determine the dimensions of the cutout of the sofa.

$$\frac{\text{length}_{\text{cutout}}}{\text{length}_{\text{actual}}} = \frac{10}{192} \qquad \frac{\text{width}_{\text{cutout}}}{\text{width}_{\text{actual}}} = \frac{10}{192}$$

$$\frac{\text{length}_{\text{cutout}}}{72 \text{ in}} = \frac{10}{192} \qquad \frac{\text{width}_{\text{cutout}}}{24 \text{ in}} = \frac{10}{192}$$

$$\text{length}_{\text{cutout}} = \frac{72 \cdot 10}{192} \qquad \text{width}_{\text{cutout}} = \frac{24 \cdot 10}{192}$$

$$\text{length}_{\text{cutout}} = 3.75 \text{ in} \qquad \text{width}_{\text{cutout}} = 1.25 \text{ in}$$

Converting decimals to fractions, the dimensions of the cutout are $3\frac{3}{4} \times 1\frac{1}{4}$. The correct choice is **(3)**.

9 a) Given $\overline{FL} \parallel \overline{IP}$, we know $\angle F \cong \angle P$ and $\angle L \cong \angle I$ because they are the alternate interior angles formed by parallel lines and a transversal. Therefore, $\triangle SIP \sim \triangle SLF$ by AA. (Note the correct order of vertices in the similarity statement.)

b) Corresponding parts are in the same ratio, so

$$\frac{PS}{FS} = \frac{IP}{LF}$$
$$= \frac{4}{6}$$
$$= \frac{2}{3}$$

c) We have to use the reciprocal of the ratio in part (b) because we want to enlarge the smaller triangle and not reduce it. A rotation of $180°$ about point S will align $\overline{SP}$ with $\overline{FS}$ and $\overline{IS}$ with $\overline{SL}$. A dilation with center at S and a scale factor of $\frac{3}{2}$ will then map $\triangle SIP$ to $\triangle SLF$.

10 $\overline{VG}$ and $\overline{TF}$ are medians that intersect at centroid K. The centroid divides $\overline{VG}$ in a 1 : 2 ratio, so

$$VK = 2KG$$
$$5x + 5 = 2(2x + 8)$$
$$5x + 5 = 4x + 16$$
$$5x = 4x + 11$$
$$x = 11$$

VG is equal to the sum of VK and KG, which can be evaluated at $x = 11$.

$$VG = VK + KG$$
$$= 5x + 5 + 2x + 8$$
$$= 7x + 13$$
$$= 7(11) + 13$$
$$= 90$$

11 $\overline{CO}$, $\overline{OW}$, and $\overline{WC}$ are all midsegments, and their lengths are $\frac{1}{2}$ the length of the opposite side of $\triangle PIG$. Therefore,

$$PI = 2 \cdot WC$$
$$= 2 \cdot 5$$
$$= 10$$

$$IG = 2 \cdot OC$$
$$= 2 \cdot 6$$
$$= 12$$

$$GP = 2 \cdot WO$$
$$= 2 \cdot 7$$
$$= 14$$

The perimeter of $\triangle PIG = PI + IG + GP$
$$= 10 + 12 + 14$$
$$= 36$$

12 An altitude to the hypotenuse of a right triangle forms 3 similar triangles. The three similar triangles are $\triangle CDB$, $\triangle BDA$, and $\triangle CBA$. Fill in the table of side lengths, and look for two pairs of corresponding sides that can be used to write a proportion involving the unknown length x.

	Leg 1	**Leg 2**	**Hypotenuse**
Small $\triangle CDB$		x	6
Medium $\triangle BDA$	x		12
Large $\triangle CBA$	6	12	

We don't have a full set of four corresponding sides, so we need to use the Pythagorean theorem on the large triangle to find the hypotenuse AC.

$$BC^2 + AB^2 = AC^2$$
$$6^2 + 12^2 = AC^2$$
$$AC^2 = 180$$
$$AC = \sqrt{180}$$
$$= 6\sqrt{5}$$

Now the table indicates two pairs of corresponding parts can be used to write a proportion.

	Leg 1	Leg 2	Hypotenuse
Small $\triangle CDB$		x	6
Medium $\triangle BDA$	x		12
Large $\triangle CBA$	6	12	$6\sqrt{5}$

$$\frac{x}{12} = \frac{6}{6\sqrt{5}}$$

$6x\sqrt{5} = 72$ cross products are equal

$$x = \frac{72}{6\sqrt{5}}$$

$$= \frac{12}{\sqrt{5}}$$

$$= \frac{12}{\sqrt{5}} \cdot \frac{\sqrt{5}}{\sqrt{5}}$$ rationalize the denominator

$$= \frac{12\sqrt{5}}{5}$$

13 The dilation tells us that the ratio of corresponding parts between rectangles $AEFG$ and $ABCD$ is $2:3$. GA and AD are corresponding parts, so the following proportion can be written:

$$\frac{GA}{AD} = \frac{2}{3}$$

$$\frac{8}{AD} = \frac{2}{3}$$

$2AD = 24$ cross products are equal

$AD = 12$

$AE = AD + DE$ segment addition

$AE = 12 + 12$

$= 24$

Now we can find AB by writing a proportion involving corresponding sides AE and AB.

$$\frac{AE}{AB} = \frac{2}{3}$$

$$3AE = 2AB \quad \text{cross products are equal}$$

$$3(24) = 2AB$$

$$72 = 2AB$$

$$AB = 36$$

Since $ABCD$ is a rectangle, opposite sides are congruent, and the perimeter is equal to

$$\text{perimeter}_{ABCD} = 2AB + 2BC$$

$$= 2(36) + 2(12)$$

$$= 72 + 24$$

$$= 96$$

14 A dilation and a reflection are required to map $ARSTUV$ to $ABCDEF$. Since R is a midpoint, we know $AR = \frac{1}{2}AB$, and a dilation with a scale factor of 2 centered at A is needed to map $\overline{AR}$ to $\overline{AB}$. The second transformation needed is a reflection over line $\overleftrightarrow{AB}$.

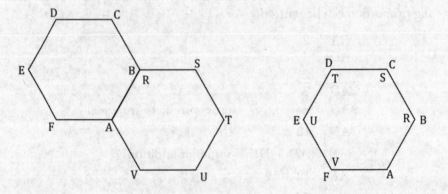

The transformation is a dilation centered at A with a scale factor of 2, followed by a reflection over $\overline{AB}$. Alternatively, the order could be reversed.

15 The strategy is to prove $\triangle TUV \sim \triangle TRS$, which leads to the proportion of corresponding sides $\dfrac{TR}{TU} = \dfrac{TS}{TV}$. Algebraic rearranging gives the final result.

Statement	Reason
1. $\overline{RUT}$, $\overline{SVT}$, and $\overline{UV} \parallel \overline{RS}$	1. Given
2. $\angle TUV \cong \angle R$ and $\angle TVU \cong \angle S$	2. Corresponding angles formed by parallel lines are congruent
3. $\triangle TUV \sim \triangle TRS$	3. AA
4. $\dfrac{TR}{TU} = \dfrac{TS}{TV}$	4. Corresponding sides in similar triangles are proportional
5. $TR = TU + UR$ $TS = TV + VS$	5. Partition
6. $\dfrac{TU + UR}{TU} = \dfrac{TV + VS}{TV}$	6. Substitution
7. $1 + \dfrac{UR}{TV} = 1 + \dfrac{VS}{TV}$	7. Division
8. $\dfrac{UR}{TU} = \dfrac{VS}{TV}$	8. Subtraction

16 The strategy is to show that $\overline{BC} \parallel \overline{DE}$ and then to use the corresponding angle relationship to show $\angle ABC \cong \angle BDE$.

Statement	Reason
1. $\overline{ABD}$, $\angle C \cong \angle E$, $\angle DEB \cong \angle EBC$	1. Given
2. $\overline{BC} \parallel \overline{DE}$	2. Two lines are parallel if the alternate interior angles formed by a transversal are congruent
3. $\angle ABC \cong \angle BDE$	3. The corresponding angles formed by two parallel lines are congruent
4. $\triangle ABC \sim \triangle BDE$	4. AA

17 The strategy is to use the theorem that an altitude to the hypotenuse of a right triangle forms three similar triangles. Two proportions from these triangles can be written and combined to yield the Pythagorean theorem.

Statement	Reason
1. Right triangle $\triangle ABC$ with a right angle at C, altitude $\overline{CD}$ drawn to hypotenuse $\overline{AC}$	1. Given
2. $\triangle BDC \sim \triangle CDA \sim \triangle BCA$	2. The altitude to the hypotenuse of a right triangle forms 3 similar right triangles
3. $\dfrac{AB}{AC} = \dfrac{AC}{AD}, \dfrac{AB}{BC} = \dfrac{BC}{BD}$	3. Corresponding sides in similar triangles are proportional
4. $(AC)^2 = AB \cdot AD,$ $(BC)^2 = AB \cdot BD$	4. The cross products of a proportion are equal
5. $(AC)^2 + (BC)^2 = AB \cdot AD + AB \cdot BD$	5. Addition property
6. $(AC)^2 + (BC)^2 = AB(AD + BD)$	6. Factor
7. $(AC)^2 + (BC)^2 = AB(AB)$	7. Partition property
8. $(AC)^2 + (BC)^2 = (AB)^2$	8. Simplify

3.8 TRIGONOMETRY

DEFINITION OF SIDES IN A RIGHT TRIANGLE RELATIVE TO AN ANGLE

- Hypotenuse—the side across from the right angle
- Opposite—the side across from the specified acute angle
- Adjacent—the side included between the right angle and the specified angle

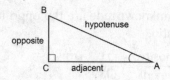 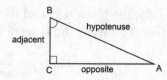

a) Opposite, adjacent, and hypotenuse relative to ∠A

b) Opposite, adjacent, and hypotenuse relative to ∠B

TRIGONOMETRIC RATIOS

Ratio	Abbreviation	Definition
Sine	Sin	$\dfrac{\text{opposite}}{\text{hypotenuse}}$
Cosine	Cos	$\dfrac{\text{adjacent}}{\text{hypotenuse}}$
Tangent	Tan	$\dfrac{\text{opposite}}{\text{adjacent}}$

The ratios sine, cosine, and tangent will have fixed values for a given acute angle in a right triangle. This is because any two right triangles with a pair of congruent acute angles are similar, and corresponding parts of similar triangles are in the same ratio. These three trigonometric ratios can be used to find a missing side of a triangle given one angle and side by setting up the correct proportion. Be sure your calculator is in degree mode if the angle is given in degrees.

Example

$\triangle SOX$ has a right angle at O. The measure of angle $S = 72°$ and $XO = 12$. Find the length OS to the nearest tenth.

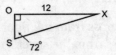

Solution:

Relative to $\angle S$, the given side and the unknown side are the opposite and adjacent, which suggest using the tangent ratio.

$$\tan(S) = \frac{\text{opposite}}{\text{adjacent}} = \frac{OX}{OS}$$

$$\tan(72°) = \frac{12}{OS}$$

$$OS = \frac{12}{\tan(72°)}$$

$$OS = \frac{12}{3.07768} = 3.89903$$

$$= 3.9$$

INVERSE TRIGONOMETRIC RATIOS

The inverse of a function "undoes" the original function. For example, $f(x) = \sqrt{x}$ and $g(x) = x^2$ are inverse functions. If $x = 5$, then $f(g(5)) = \sqrt{5^2} = 5$. The trigonometric functions have the following inverse functions:

$\sin(x)$	$\arcsin(x)$
$\cos(x)$	$\arccos(x)$
$\tan(x)$	$\arctan(x)$

- These three inverse functions are often abbreviated $\sin^{-1}(x)$, $\cos^{-1}(x)$, and $\tan^{-1}(x)$. They can be used to find the measure of an angle given a ratio of any two sides in the triangle.

Example

Using the accompanying figure, find the measure of angles M and N.

Solution:

$$\cos(M) = \frac{3}{7}$$

$$m\angle M = \cos^{-1}\left(\frac{3}{7}\right)$$

$$= 64.6°$$

$$\sin(N) = \frac{3}{7}$$

$$m\angle N = \sin^{-1}\left(\frac{3}{7}\right)$$

$$= 25.4°$$

COFUNCTIONS

Cofunction Relationship

The sine of an angle is equal to the cosine of its complement: $\sin(A) = \cos(90 - A)$.

The cosine of an angle is equal to the sine of its complement: $\cos(A) = \sin(90 - A)$.

$$\sin(A) = \cos(B) = \cos(90 - A) = \frac{BC}{AB}$$

$$\sin(B) = \cos(A) = \sin(90 - A) = \frac{AC}{AB}$$

In other words, the cofunction relationship states that if a sine is equal to a cosine, then the two angles must sum to 90°.

Example

Given $\sin(2x + 4) = \cos(3x + 21)$. Solve for x.

Solution:

The cofunctions are equal, so the angles must be complementary.

$$
\begin{aligned}
2x + 4 + 3x + 21 &= 90 \\
5x + 25 &= 90 \\
5x &= 65 \\
x &= 13
\end{aligned}
$$

MODELING WITH TRIGONOMETRY

- *Angle of elevation*—The angle formed by the horizontal and the line directed upward to an object
- *Angle of depression*—The angle formed by the horizontal and the line directed downward to an object

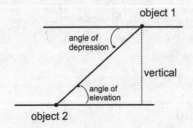

Strategy for trigonometry modeling problems:

- Start with a detailed sketch.
- Identify the triangles formed and the relevant trigonometric ratios.
- Consider parallel lines and other relationships if you are still missing parts. Notice that the angle of elevation and angle of depression are congruent alternate interior angles.
- Consider working with more than one triangle, especially if the problem involves an object that moves from one position to another.

SINE FORMULA FOR AREA OF A TRIANGLE

There are two formulas for the area of a triangle. The first is the familiar area $= \frac{1}{2}$base $\cdot$ height. The second is a related version that uses the length of any two sides and the sine of the included angle.

Formula	Figure
area $= \frac{1}{2}$base $\cdot$ height	
area $= \frac{1}{2}a \cdot b \cdot \sin C$ area $= \frac{1}{2}b \cdot c \cdot \sin A$ area $= \frac{1}{2}a \cdot c \cdot \sin B$	

SPECIAL RIGHT TRIANGLES

You can find a missing side of any 30°-60°-90° triangle or any 45°-45°-90° triangle by using known side lengths and creating a proportion.

Triangle	Figure
30°-60°-90°	
45°-45°-90°	

Example

Find the lengths of *RP* and *QP* in triangle *RQP*.
Express your answer in simplest radical form.

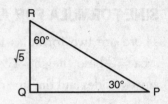

Solution:

Triangle *RQP* is a 30°-60°-90° triangle, the known side *RQ* is across from the 30° angle, and the unknown side *QP* is across from the 60° angle. Based on the table above, we know the ratio *RQ* : *QP* must be $1 : \sqrt{3}$. In the same way, we know the ratio *RQ* : *RP* must be 1 : 2.

Find **QP**	Find **RP**
$\dfrac{\sqrt{5}}{QP} = \dfrac{1}{\sqrt{3}}$ $QP = \sqrt{3} \cdot \sqrt{5}$ $QP = \sqrt{15}$	$\dfrac{\sqrt{5}}{RP} = \dfrac{1}{2}$ $RP = 2\sqrt{5}$

Practice Exercises

1 In $\triangle ABC$, $m\angle B = 90°$, $m\angle C = 51°$, and $BC = 13$. What is the length AB?

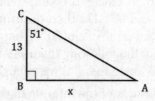

(1) 16.1   (3) 18.2

(2) 16.7   (4) 20.7

2 $\triangle JPL$ is a right triangle with a right angle at L. If $m\angle J = 19°$ and $\overline{LJ} = 21$, what is the length of $\overline{PJ}$?

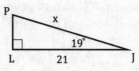

(1) 21.5   (3) 61.0

(2) 22.2   (4) 64.5

3 A 24 ft ladder is leaning against a wall. The base of the ladder is 5 ft from the house. What angle does the ladder make with the ground?

   (1) 12°   (3) 67°

   (2) 65°   (4) 78°

4 In right triangle PQR, $m\angle Q = 90°$. If $\sin(P) = 4x^2 + 5x + 2$ and $\cos(R) = 5x + 3$, what is the measure of $\angle P$?

   (1) 5°   (3) 25°

   (2) 8°   (4) 30°

5 John constructs three right triangles of different sizes, each having angles that measure 30°, 60°, and 90°. He then measures the side lengths of each triangle and calculates the value of sin(60°) from the side lengths using each of the triangles. Which of the following theorems would best explain why the value he calculates is the same for all three triangles?

   (1) The sum of the measures of the interior angles of a triangle equals 180°.

   (2) Corresponding parts of congruent triangles are congruent.

   (3) Corresponding sides of similar triangles are proportional.

   (4) The longest side of a triangle is always opposite the largest angle.

6 The side of Rayquan's house is 30 ft long. Rayquan is planning to install a gutter along the roof as shown. He wants the angle of depression to be 3° so the water will drain out at the end of the gutter. How long should the gutter be?

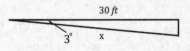

   (1) 31 ft 7 in   (3) 30 ft 4 in

   (2) 31 ft 2 in   (4) 30 ft 0.5 in

7 In $\triangle ABC$, m$\angle B = 90°$, $\tan(A) = m$, and $\sin(C) = \dfrac{1}{p}$. What is the value of $\cos(C)$?

(1) $\dfrac{m}{p}$ (3) $\dfrac{p}{m}$

(2) $m \cdot p$ (4) $m + p$

8 In $\triangle PDX$, m$\angle P = 28°$, $PD = 8$, and $PX = 18$. Find the area of $\triangle PDX$, rounded to the nearest unit.

(1) 17   (3) 68
(2) 34   (4) 127

9 In $\triangle QRS$, $QS = 6$, m$\angle Q = 30°$, and m$\angle S = 60°$. Find the length of side RQ.

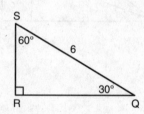

10 Jack is watching the launch of a rocket from a viewing area 4,500 feet from the launch pad. At 15 seconds after launch, he measures a 67° angle of elevation from the ground to the rocket. What is the average speed of the rocket during the first 15 seconds of its flight? Assume the rocket travels upward, perpendicular to the ground, and give your answer to the nearest foot per second.

11 Rick is flying a kite in the park. He holds the spool of string 3 ft above the ground, and lets out 200 ft of string. The kite is initially flying with a 72° angle of elevation.

a) What is the altitude of the kite?

b) The wind shifts and the angle of elevation altitude decreases to 52°. How many feet does the kite drop?

12 Frank is standing at the top of a water tower and looks down at the town below. Frank measures an angle of depression of 52° to his house, and he knows the water tower is 120 ft tall. How far along the ground is the tower from his house? Round to the nearest foot.

13 A lighthouse sits on a 100 ft cliff. An observer in a boat in the harbor notes an angle of elevation to the base of the cliff of 28°, and an angle of elevation to the top of the lighthouse of 44°. What is the height of the lighthouse, rounded to the nearest foot?

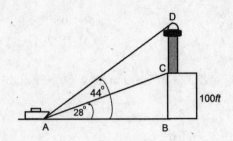

Solutions

1 Relative to $\angle C$, $\overline{AB}$ is the opposite and $\overline{BC}$ is the adjacent, so the tangent ratio applies.

$$\text{tangent} = \frac{\text{opposite}}{\text{adjacent}}$$

$$\tan(C) = \frac{AB}{BC}$$

$$\tan(51°) = \frac{x}{13}$$

$$x = 13\tan(51°)$$

$$= 16.0536$$

$$= 16.1$$

The correct choice is **(1)**.

2 Relative to $\angle J$, $\overline{LJ}$ is the adjacent and $\overline{JP}$ is the hypotenuse. The cosine ratio applies here.

$$\text{cosine} = \frac{\text{adjacent}}{\text{hypotenuse}}$$

$$\cos(J) = \frac{LJ}{JP}$$

$$\cos(19°) = \frac{21}{x}$$

$$x = \frac{21}{\cos(19°)}$$

$$= 22.2100$$

$$= 22.2$$

The correct choice is **(2)**.

3 The ladder, wall, and ground form a right triangle as shown.

Relative to the desired angle, the 5 ft distance is the adjacent and the 24 ft length is the hypotenuse. Adjacent and hypotenuse suggest using the cosine. Since we are looking for the angle, apply the inverse cosine function.

$$\cos(x) = \frac{\text{adjacent}}{\text{hypotenuse}}$$

$$\cos(x) = \frac{5}{24}$$

$$x = \cos^{-1}\left(\frac{5}{24}\right)$$

$$= 77.9753°$$

$$= 78°$$

The correct choice is **(4)**.

4 The sine of an angle and the cosine of its complement are always equal. Since $\triangle PQR$ is a right triangle with the right angle at Q, angles P and R must be complementary. Setting $\sin(P)$ and $\cos(R)$ equal, we get:

$$\sin(P) = \cos(R)$$

$$4x^2 + 5x + 2 = 5x + 3$$

$$4x^2 = 1$$

$$x^2 = \frac{1}{4}$$

$$x = \pm\frac{1}{2}$$

Substituting $x = -\frac{1}{2}$ into the expression for $\sin(P)$, we get

$$\sin(P) = 4x^2 + 5x + 2$$

$$= 4\left(-\frac{1}{2}\right)^2 + 5\left(-\frac{1}{2}\right) + 2$$

$$= 1 - 2\frac{1}{2} + 2$$

$$= \frac{1}{2}$$

Now substitute $x = \dfrac{1}{2}$.

$$\sin(P) = 4\left(\dfrac{1}{2}\right)^2 + 5\left(\dfrac{1}{2}\right) + 2$$

$$= 1 + 2\dfrac{1}{2} + 2$$

$$= 5\dfrac{1}{2}$$

The sine function cannot be greater than 1, so $x = \dfrac{1}{2}$ is not a solution. Using the solution $x = -\dfrac{1}{2}$ and $\sin(P) = \dfrac{1}{2}$, find m$\angle P$ using the inverse sine function.

$$\sin(P) = \dfrac{1}{2}$$

$$m\angle P = \sin^{-1}\left(\dfrac{1}{2}\right)$$

$$m\angle P = 30°$$

The correct choice is (**4**).

5 The three triangles are all similar by the angle–angle theorem. The sine ratio is equal to the $\dfrac{\text{opposite}}{\text{hypotenuse}}$. Since all pairs of corresponding sides in similar triangles are proportional, the ratio $\dfrac{\text{opposite}}{\text{hypotenuse}}$ will be equal to the same value in all three triangles.

The correct choice is (**3**).

6 Relative to the 3° angle, the horizontal distance is the adjacent and the gutter is the hypotenuse, so we can apply the cosine ratio.

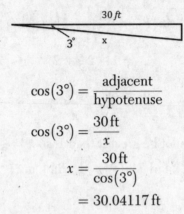

$$\cos(3°) = \frac{\text{adjacent}}{\text{hypotenuse}}$$

$$\cos(3°) = \frac{30\,\text{ft}}{x}$$

$$x = \frac{30\,\text{ft}}{\cos(3°)}$$

$$= 30.04117\,\text{ft}$$

To convert to feet and inches, multiply the decimal part of the number by 12.

$$0.4117\,(12) = 0.49405\,\text{in}$$

The gutter length should be 30 ft 0.49405 in. Rounded to the nearest 0.1 inch, the length is 30 ft 0.5 in.

The correct choice is **(4)**.

7 Making a sketch of △ABC, we can fill in the lengths of AB, BC, and AC as shown in the figure.

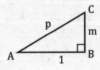

The tangent is equal to $\dfrac{\text{opposite}}{\text{adjacent}}$, so

$$\tan(A) = m$$

$$\frac{BC}{AB} = \frac{m}{1}$$

Since we are only concerned with ratios, we can assume $BC = m$ and $AB = 1$.

The second ratio states

$$\sin(C) = \frac{1}{p}$$

$$\frac{AB}{AC} = \frac{1}{p}$$

Since we assumed $AB = 1$, we can assume $AC = p$. Cos(C) can now be calculated.

$$\cos(C) = \frac{\text{adjacent}}{\text{hypotenuse}}$$

$$= \frac{BC}{AC}$$

$$= \frac{m}{p}$$

The correct choice is **(1)**.

8 The two sides that form angle $\angle P$ are $\overline{PD}$ and $\overline{PX}$. Apply the area formula for a triangle.

$$\text{area} = \frac{1}{2}a \cdot b \cdot \sin C$$

$$\text{area} = \frac{1}{2} \cdot 8 \cdot 18 \cdot \sin 28° = 33.8$$

The correct choice is **(2)**.

9 RQ is opposite the 60° angle, and the known side SQ is opposite the right angle. Use the standard 30°-60°-90° triangle to set up a proportion and solve.

$$\frac{RQ}{SQ} = \frac{\sqrt{3}}{2}$$

$$\frac{RQ}{6} = \frac{\sqrt{3}}{2}$$

$$2RQ = 6\sqrt{3}$$

$$RQ = 3\sqrt{3}$$

10 The path of the rocket traveling upward forms a right triangle with the ground. The distance to the rocket is the adjacent, and the height of the rocket is the opposite, so the tangent ratio can be applied.

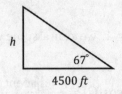

$$\tan(67°) = \frac{h}{4,500}$$

$$h = 4,500\tan(67°)$$

$$= 4,500(2.3558)$$

$$= 10,601.33 \text{ ft}$$

To find the speed, apply the relationship rate $= \dfrac{\text{distance}}{\text{time}}$.

$$\text{speed} = \frac{10,601.33\,\text{ft}}{15\,\text{sec}}$$

$$= 706.73\,\text{ft/sec}$$

$$= 707\,\text{ft/sec}$$

11 First make a sketch. A right triangle is formed by the kite string, the altitude measured from the spool (x), and the horizontal distance to the kite. The total altitude will be $(x + 3)$ ft. To find how many feet the kite drops, calculate the altitude at the two angles of elevation, and then find the difference.

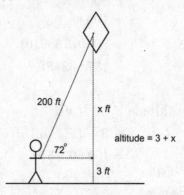

a) Relative to the angle of elevation, the 200 ft kite string is the hypotenuse, and the altitude measured from the spool is opposite. The opposite and hypotenuse suggest the sine ratio. Find x using the following:

$$\sin(72°) = \frac{\text{opposite}}{\text{hypotenuse}}$$

$$\sin(72°) = \frac{x}{200}$$

$$x = 200\sin(72°)$$

$$x = 200(0.951056)$$

$$= 190.2112\,\text{ft}$$

Next, calculate the altitude from x.

$$\text{Altitude} = 3\,\text{ft} + x\,\text{ft}$$

$$= 3\,\text{ft} + 190.21123\,\text{ft}$$

$$= 193.21123\,\text{ft}$$

$$= 193\,\text{ft}$$

b) We can use the same equations to calculate the new altitude if the angle of elevation decreases to 52°.

$$\sin(52°) = \frac{x}{200}$$
$$x = 200\sin(52°)$$
$$x = 200(0.788010)$$
$$= 157.60215\,\text{ft}$$

$$\text{Altitude} = 3\,\text{ft} + x\,\text{ft}$$
$$= 3 + 157.60215\,\text{ft}$$
$$= 160.60215\,\text{ft}$$

The change in altitude is

$$193.21123\ \text{ft} - 160.60215\ \text{ft}$$
$$= 32.6090\ \text{ft}$$
$$= 33\ \text{ft}$$

The kite drops 33 ft.

12 Start with a sketch, and fill in the appropriate dimensions, horizontals, and verticals.

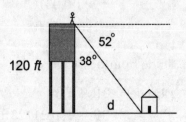

The angle of depression is 52°. It is easier in the problem to find the angle complementary to the angle of depression and work with that triangle. The complementary angle is 38°. The height of the tower is adjacent to the 38° angle, and the distance along the ground is opposite from the 38° angle, so we use the tangent ratio.

$$\tan(38°) = \frac{d}{120\,\text{ft}}$$

$$d = 120\,\text{ft} \cdot \tan(38°)$$

$$d = 93.75\,\text{ft}$$

The house is 94 ft from the water tower.

13 First find length AB using $\triangle ABC$; then find length BD using $\triangle ABD$.

In $\triangle ABC$, BC is opposite $\angle A$, and AB is the adjacent, so we can use the tangent.

$$\tan = \frac{\text{opposite}}{\text{adjacent}}$$

$$\tan(\angle BAC) = \frac{BC}{AB}$$

$$\tan(28°) = \frac{100}{AB}$$

$$AB = \frac{100}{\tan(28°)}$$

$$AB = 188.0726$$

In $\triangle ABD$, AB is the adjacent and BD is the opposite, so we use the tangent again.

$$\tan(\angle BAD) = \frac{BD}{AD}$$

$$\tan(44°) = \frac{BD}{188.0726}$$

$$BD = 188.0726\tan(44°)$$

$$BD = 181.6196$$

The height of the lighthouse, DC, is found by subtracting the height of the cliff from BD.

$$DC = BD - BC$$

$$DC = 181.6196 - 100$$

$$DC = 81.6196$$

The height of the lighthouse is 82 ft.

3.9 PARALLELOGRAMS

PROPERTIES OF PARALLELOGRAMS

A **parallelogram** is a quadrilateral whose opposite sides are parallel. All parallelograms have the following properties:

1. Opposite sides are congruent.

2. Opposite angles are congruent.

3. Adjacent angles are supplementary.

4. The diagonals bisect each other.

5. The two diagonals each divide the parallelogram into two congruent triangles.

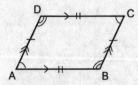 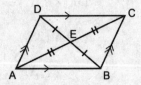

Congruent sides:
$AB = CD, BC = AD$

Congruent angles:
$\angle A \cong \angle C, \angle B \cong \angle D$

Supplementary angles:
$\angle A$ and $\angle B$, $\angle B$ and $\angle C$
$\angle C$ and $\angle D$, $\angle D$ and $\angle A$

Diagonals bisect each other:
$AE = EC, BE = ED$

Diagonals form congruent triangles:
$\triangle BAD \cong \triangle DCB, \triangle ADC \cong \triangle CBA$

SPECIAL PARALLELOGRAMS

Rectangle—a parallelogram with right angles
Rhombus—a parallelogram with consecutive congruent sides
Square—a parallelogram with right angles and consecutive congruent sides

Rectangles, rhombuses, and squares have all the properties of parallelograms plus the following properties:

	Rectangle	Rhombus	Square
4 right angles	✓		✓
Congruent diagonals	✓		✓
4 congruent sides		✓	✓
Perpendicular diagonals		✓	✓
Diagonals bisect the angles		✓	✓

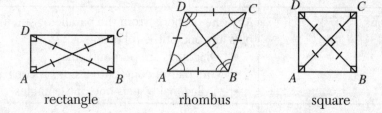

rectangle   rhombus   square

TRAPEZOIDS

A *trapezoid* is a quadrilateral with exactly one pair of parallel sides. A trapezoid with one pair of parallel sides is shown in the accompanying figure. The parallel sides are called the bases and the nonparallel sides are called the legs. The same side interior angles formed with the parallel bases are supplementary.

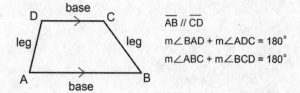

$\overline{AB} \parallel \overline{CD}$

$m\angle BAD + m\angle ADC = 180°$

$m\angle ABC + m\angle BCD = 180°$

CLASSIFYING PARALLELOGRAMS AND TRAPEZOIDS

The following table lists the properties sufficient to prove a quadrilateral is a parallelogram, rectangle, rhombus, square, or trapezoid.

Classification	What's Needed to Classify the Figure
Trapezoid	One pair of parallel sides
Parallelogram	Any *one* of the following: • Two pairs of opposite sides parallel • Two pairs of opposite sides congruent • Two pairs of opposite angles congruent • Consecutive angles supplementary • Diagonals that bisect each other • One pair of sides congruent *and* parallel
Rectangle	Any one property from the parallelogram list, plus any one of the following: • One right angle • Diagonals that are congruent
Rhombus	Any one property from the parallelogram list, plus any one of the following: • Diagonals are perpendicular • One pair of consecutive sides is congruent • A diagonal bisects one of the angles
Square	Any one property from the parallelogram list *plus* Any one property from the rectangle list *plus* Any one property from the rhombus list

The parallelogram properties needed for a proof are illustrated in the following sketches.

Property of Parallelogram	What It Looks Like
Two pairs of opposite sides are parallel.	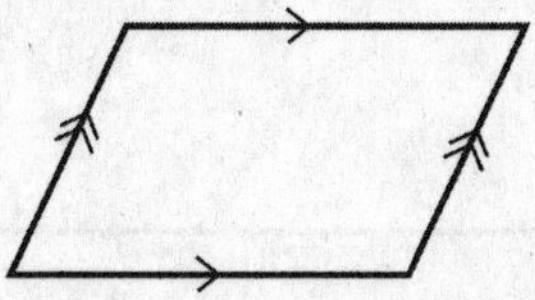
Two pairs of opposite sides are congruent.	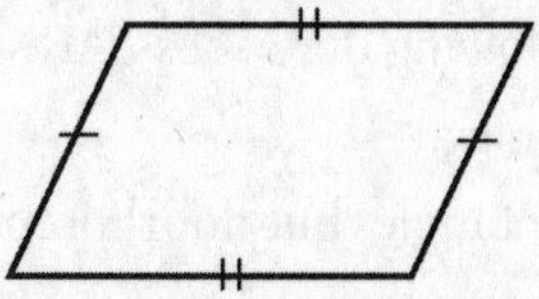
One pair of parallel sides is congruent and parallel.	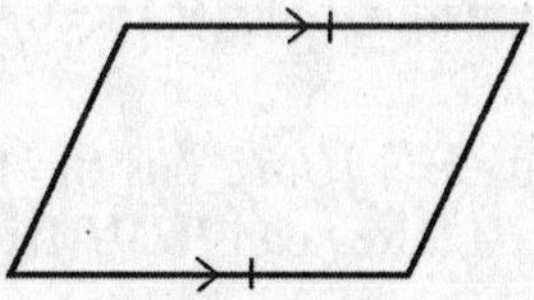
Two pairs of opposite angles are congruent.	
Two pairs of consecutive angles are supplementary.	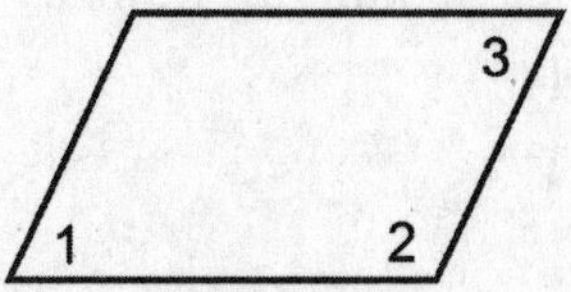 ∠1 and ∠2, ∠2 and ∠3 are supplementary
The diagonals bisect each other.	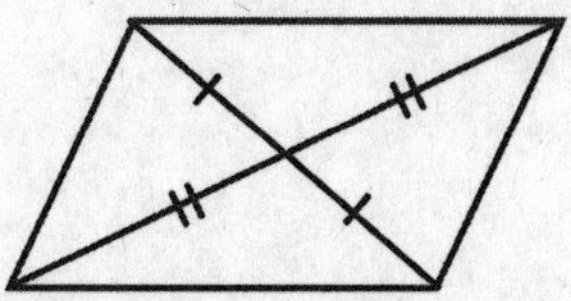

148

Practice Exercises

1 Parallelogram *PARK* has congruent and perpendicular diagonals. Which of the following describes *PARK*?

(1) square

(2) rectangle, but not rhombus

(3) rhombus, but not rectangle

(4) trapezoid, but not rectangle or rhombus

2 Quadrilateral *JUMP* has the following properties: $JU = MP$, $UM = PJ$, and $JU = UM$. How can *JUMP* be classified?

(1) parallelogram only

(2) parallelogram and rectangle

(3) parallelogram and rhombus

(4) not enough information to make any classification

3 In parallelogram *RSTU*, $m\angle R = 8x + 12$ and $m\angle T = 4x + 24$. What is $m\angle R$?

(1) 144°

(2) 72°

(3) 36°

(4) 18°

4 In parallelogram *ABCD*, the measures of $\angle B$ and $\angle C$ are in a 3 : 2 ratio. What is the measure of $\angle B$?

(1) 120°

(2) 108°

(3) 72°

(4) 60°

5 What are the measures of ∠1 and ∠2 in the given trapezoid?

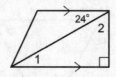

(1) m∠1 = 24° and m∠2 = 24°

(2) m∠1 = 66° and m∠2 = 24°

(3) m∠1 = 66° and m∠2 = 66°

(4) m∠1 = 24° and m∠2 = 66°

6 Given rhombus $ABCD$ with diagonals $\overline{AC}$ and $\overline{BD}$ intersecting at E. If m∠DAE = 41°, what is the measure of ∠CBE?

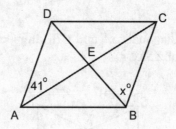

(1) 20.5°   (3) 45°

(2) 41°   (4) 49°

7 $ABCD$ is a rhombus with diagonals $\overline{AC}$ and $\overline{BD}$ intersecting at E. If $AC = 16$ and $BD = 12$, what is the length of $\overline{AB}$?

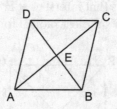

(1) 14   (3) 10

(2) $4\sqrt{14}$ (4) 28

8 Find the measure of ∠C in the parallelogram $ABCD$.

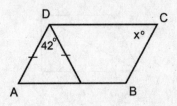

(1) 42°   (3) 69°

(2) 58°   (4) 71°

9 Diagonal $\overline{QS}$ is drawn in rectangle $QRST$. If m∠RQS = $(4x + 2)°$ and m∠QSR = $(3x + 11)°$, what is m∠RQS?

(1) 11°   (3) 46°

(2) 44°   (4) 90°

10 In rectangle $CARS$, diagonals $\overline{SA}$ and $\overline{CR}$ intersect at X. If m∠RSX = 28°, what is the measure of ∠SXC?

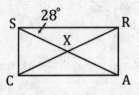

(1) 56°   (3) 66°

(2) 62°   (4) 68°

11 Square $STAR$ has diagonals that intersect at P. If the perimeter of $STAR$ is 24, what is the length of RP in simplest radical form?

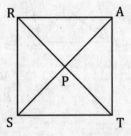

12 Given: $\angle E \cong \angle G$ and $\angle EDF \cong \angle GFD$

Prove: $DEFG$ is a parallelogram

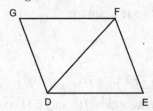

13 Given: $\overline{DKB}$, $\overline{DLC}$, $PLDK$ is a rhombus, and $\overline{BK} \cong \overline{CL}$

Prove: $\angle B \cong \angle C$

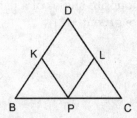

14 Given: Rectangle $ABCD$ with diagonals $\overline{AC}$ and $\overline{BD}$

Prove: The diagonals of a rectangle are congruent $\left(\overline{AC} \cong \overline{BD}\right)$

15 Given: $\angle BCA \cong \angle DAC$, $\angle BAC \cong \angle BCA$, $\overline{BC} \cong \overline{AD}$

Prove: $ABCD$ is a rhombus

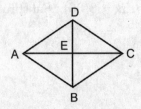

Solutions

1 A parallelogram with perpendicular diagonals is a rhombus, and a parallelogram with congruent diagonals is a rectangle. A figure with the properties of a rectangle and a rhombus is a square.

The correct choice is **(1)**.

2 Opposites sides $\overline{JU}$ and $\overline{MP}$, and $\overline{UM}$ and $\overline{PJ}$ are congruent, making *JUMP* a parallelogram. Consecutive sides $\overline{JU}$ and $\overline{UM}$ are congruent, making *JUMP* a rhombus.

The correct choice is **(3)**.

3    $\text{m}\angle R = \text{m}\angle T$ Opposite angles of a parallelogram are congruent

$$8x + 12 = 4x + 24$$
$$4x + 12 = 24$$
$$4x = 12$$
$$x = 3$$
$$\text{m}\angle R = 8(3) + 12 \quad \text{Substitute } x = 3$$
$$= 24 + 12$$
$$= 36°$$

The correct choice is **(3)**.

4 Let $\text{m}\angle B = 3x$ and $\text{m}\angle C = 2x$, and apply the parallelogram property that consecutive angles are supplementary.

$$\text{m}\angle B + \text{m}\angle C = 180°$$
$$3x + 2x = 180$$
$$5x = 180$$
$$x = 36$$
$$\text{m}\angle B = 3(36) \quad \text{Substitute } x = 36$$
$$= 108°$$

The correct choice is **(2)**.

5 The diagonal is a transversal forming a pair of congruent alternate interior angles, so m$\angle 1 = 24°$. The triangle angle sum theorem can be used on the bottom triangle to find $\angle 2$.

$$\text{m}\angle 1 + \text{m}\angle 2 + 90° = 180°$$
$$24° + \text{m}\angle 2 + 90° = 180°$$
$$\text{m}\angle 2 + 114° = 180°$$
$$\text{m}\angle 2 = 66°$$

The correct choice is **(4)**.

6 m$\angle BCE = 41°$ $\angle DAE$ and $\angle BEC$ are congruent alternate interior angles

   m$\angle BEC = 90°$ Diagonals of a rhombus are perpendicular

   m$\angle CBE + \text{m}\angle BEC + \text{m}\angle BCE = 180°$ Triangle angle sum theorem

$$\text{m}\angle CBE + 90° + 41° = 180°$$
$$\text{m}\angle CBE + 131° = 180°$$
$$\text{m}\angle CBE = 49°$$

The correct choice is **(4)**.

7 Use the rhombus property that the diagonals bisect each other and are perpendicular.

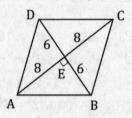

$$AE = \frac{1}{2}AC \quad \text{Diagonals of a rhombus bisect}$$
$$\text{each other}$$

$$= \frac{1}{2}(16)$$

$$= 8$$

$$BE = \frac{1}{2}BD$$

$$= \frac{1}{2}(12)$$

$$= 6$$

$\triangle AEB$ is a right triangle   Diagonals of a rhombus are
perpendicular

$$(AE)^2 + (BE)^2 = (AB)^2 \quad \text{Pythagorean theorem}$$

$$6^2 + 8^2 = (AB)^2$$

$$100 = (AB)^2$$

$$AB = 10$$

The correct choice is (**3**).

8 $m\angle A = \dfrac{180° - 42°}{2}$ Isosceles triangle theorem

$\quad = 69°$

$m\angle C = m\angle A$ Opposite angles are congruent

$\quad = 69°$

The correct choice is (**3**).

9 Start by sketching the figure.

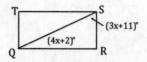

$\triangle QSR$ is a right triangle with a right angle at $\angle R$, so $\angle RQS$ and $\angle QSR$ must be complementary.

$$m\angle RQS + m\angle QSR = 90°$$
$$4x + 2 + 3x + 11 = 90$$
$$7x + 13 = 90$$
$$7x = 77$$
$$x = 11$$
$$m\angle RQS = 4(11) + 2 \quad \text{Substitute } x = 11$$
$$m\angle RQS = 46°$$

The correct choice is (**3**).

10 $\quad m\angle CSX + m\angle RSX = 90°$ Angles in a rectangle are right
angles
$\quad\quad\quad m\angle CSX + 28° = 90°$
$\quad\quad\quad\quad\quad m\angle CSX = 62°$

The diagonals of a rectangle are congruent and bisect each other; therefore, $\overline{SX} \cong \overline{CX}$, making $\triangle SXC$ isosceles.

$$m\angle SCX = m\angle CSX \quad \text{Base angles of an isosceles triangle are congruent}$$
$$= 62°$$
$$m\angle CSX + m\angle SCX + m\angle SXC = 180° \quad \text{Triangle angle sum theorem}$$
$$62° + 62° + m\angle SXC = 180°$$
$$124° + m\angle SXC = 180°$$
$$m\angle SXC = 56°$$

The correct choice is (**1**).

11   $RS = \dfrac{1}{4}$ perimeter   All sides of a square are congruent

$$= \frac{1}{4}(24)$$

$$= 6$$

$$RP = PS \qquad \text{Diagonals of a square bisect each} \\ \text{other and are congruent}$$

$$(RP)^2 + (PS)^2 = (RS)^2 \quad \text{Pythagorean theorem}$$

$$(RP)^2 + (RP)^2 = 6^2$$

$$2(RP)^2 = 36$$

$$(RP)^2 = 18$$

$$RP = \sqrt{18}$$

$$= 3\sqrt{2}$$

12

Statement	Reason
1. $\angle EDF \cong \angle GFD$	1. Given
2. $\overline{ED} \parallel \overline{FG}$	2. Two lines are parallel if the alternate interior angles formed are congruent
3. $\angle E \cong \angle G$	3. Given
4. $\overline{FD} \cong \overline{FD}$	4. Reflexive property
5. $\triangle EDF \cong \triangle GFD$	5. AAS
6. $\angle EFD \cong \angle GDF$	6. CPCTC
7. $\overline{GD} \parallel \overline{FE}$	7. Two lines are parallel if the alternate interior angles formed are congruent
8. $DEFG$ is a parallelogram	8. A quadrilateral with two pairs of opposite parallel sides is a parallelogram

13

Statement	Reason
1. $\overline{DKB}$, $\overline{DLC}$, and $PLDK$ is a rhombus	1. Given
2. $\overline{KD} \cong \overline{LD}$	2. Consecutive sides of a rhombus are congruent
3. $\overline{BK} \cong \overline{CL}$	3. Given
4. $\overline{BK} + \overline{KD} \cong \overline{CL} + \overline{LD}$	4. Addition postulate
5. $\overline{BD} \cong \overline{CD}$	5. Partition property
6. $\angle B \cong \angle C$	6. Angles opposite congruent sides in triangles are congruent

14 The strategy is to prove $\triangle ABC \cong \triangle BAD$ and then apply CPCTC.

Statement	Reason
1. Rectangle $ABCD$ with diagonals $\overline{AC}$ and $\overline{BD}$	1. Given
2. $\angle ABC$ and $\angle BAD$ are right angles	2. All angles in a rectangle are right angles
3. $\angle ABC \cong \angle BAD$	3. Right angles are congruent
4. $\overline{AD} \cong \overline{BC}$	4. Opposite sides of a rectangle are congruent
5. $\overline{AB} \cong \overline{AB}$	5. Reflexive property
6. $\triangle ABC \cong \triangle BAD$	6. SAS
7. $\overline{AC} \cong \overline{BD}$	7. CPCTC

15 The strategy is to first prove $ABCD$ is a parallelogram and then prove $\triangle ABC$ is an isosceles triangle, with legs $\overline{AB} \cong \overline{BC}$.

Statement	Reason
1. $\angle BCA \cong \angle DAC$	1. Given
2. $\overline{BC} \parallel \overline{AD}$	2. Two lines are parallel if the alternate interior angles formed by a transversal are congruent
3. $\overline{BC} \cong \overline{AD}$	3. Given
4. $ABCD$ is a parallelogram	4. A quadrilateral is a parallelogram if one pair of sides is parallel and congruent
5. $\angle BAC \cong \angle BCA$	5. Given
6. $\triangle ABC$ is isosceles	6. A triangle with congruent base angles is isosceles
7. $\overline{AB} \cong \overline{BC}$	7. Sides opposite congruent angles in a triangle are congruent
8. $ABCD$ is a rhombus	8. A parallelogram with a pair of consecutive congruent sides is a rhombus

3.10 COORDINATE GEOMETRY PROOFS

Distance, midpoint, and slope are the tools used on the coordinate plane to prove properties of a figure. They are used as follows:

Quantity	Formula	Use
Slope	$m = \dfrac{y_2 - y_1}{x_2 - x_1}$ or $\dfrac{\text{rise}}{\text{run}}$	Prove segments or lines are parallel (slopes are equal) or perpendicular (slopes are negative reciprocals)
Midpoint	$x_{\text{MP}} = \dfrac{1}{2}(x_1 + x_2)$ $y_{\text{MP}} = \dfrac{1}{2}(y_1 + y_2)$	Prove segments bisect each other (midpoints are concurrent)
Length	$d = \sqrt{(x_2 - x_1)^2 + (y_2 - y_1)^2}$	Prove segments are congruent (distances between endpoints are equal)

The steps to writing a good coordinate geometry proof are

- *Graph the points*—The graph will help you plan a strategy, check your work, and help with some calculations.
- *Plan a strategy*—Determine what property you will demonstrate and what calculations are needed.
- *Perform the calculations*—Write down the general equations you are using and clearly label which segments correspond to which calculations.
- *Write a summary statement*—You must state in words a justification that explains why your calculations justify the proof.

PROVING PARALLELOGRAMS ON THE COORDINATE PLANE

A given quadrilateral can be proven to be a parallelogram by demonstrating one of the parallelogram properties. The three properties that can be easily proven are parallel sides, congruent sides, and diagonals bisecting each other. The calculations required and suggested summary statements are shown in the following table.

Property	Calculations	Summary Statement
2 pairs of opposite sides are parallel	Slope of each side (4 slope calculations)	The slopes of opposite sides are equal; therefore, both pairs of opposite sides are parallel. A quadrilateral with two pairs of opposite sides parallel is a parallelogram.
2 pairs of opposite sides are congruent	Length of each side (4 distance calculations)	The lengths of opposite sides are equal; therefore, both pairs of opposite sides are congruent. A quadrilateral with two pairs of opposite congruent sides is a parallelogram.
The diagonals bisect each other	Midpoints of the diagonals (2 midpoint calculations)	The diagonals have the same midpoint; therefore, they bisect each other. A quadrilateral whose diagonals bisect each other is a parallelogram.

PROVING SPECIAL PARALLELOGRAMS ON THE COORDINATE PLANE

To prove a quadrilateral is a special parallelogram, a good strategy is to first prove the figure is a parallelogram and then to demonstrate one of the special properties shown in the table. For squares, you need to show that the figure is both a rectangle and a rhombus.

First show the figure is a parallelogram, then...

Figure	Property	Calculation	Example of Summary Statement
Rhombus	Perpendicular diagonals	Slope of each diagonal	The slopes of the diagonals are negative reciprocals; therefore, they are ⊥. A parallelogram with ⊥ diagonals is a rhombus.
	Two consecutive congruent sides	Length of two consecutive sides	The lengths of 2 consecutive sides are equal, so they are ≅. A parallelogram with consecutive ≅ sides is a rhombus.
Rectangle	A right angle	Slope of two consecutive sides	The slopes of two consecutive sides are negative reciprocals, so the sides are ⊥ and form a right angle. A parallelogram with a right angle is a rectangle.
	Congruent diagonals	Length of each diagonal	The lengths of the diagonals are ≅, so they are congruent. A parallelogram with congruent diagonals is a rectangle.
Square	Show the figure is both a rectangle and a rhombus.		A figure that is both a rectangle and a rhombus is a square.

One strategy to aid in remembering which properties to prove is to work only with the diagonals:

- Midpoints for a parallelogram
- Midpoints and slope for a rhombus
- Midpoints and length for a rectangle

OTHER COORDINATE GEOMETRY PROOFS

A quadrilateral can be shown to be a trapezoid using slope calculations to show two sides are parallel.

Triangles can be classified using coordinate geometry calculations:

Isosceles triangle—distance calculation to show two sides are congruent
Equilateral—distance calculation to show three sides are congruent
Right triangle—slope calculation to show two sides are perpendicular

Coordinate geometry can be used to prove special segments in triangles. The following table summarizes some of the segments and points of concurrency, and the calculations needed.

Segment	Calculation Needed	Summary Statement
Median	Midpoint	The segment has endpoints at a vertex and midpoint, so it is a median.
Altitude	Two slopes	The slopes are negative reciprocals, so the segment is ⊥ to the opposite side, making it an altitude.
Perpendicular bisector	Slope and midpoint	The slopes are negative reciprocals, making the segment ⊥, and the segment passes through the midpoint, making it a bisector. Therefore, it is a perpendicular bisector.
Midsegment	Two midpoints	The segment's endpoints are the midpoints of sides of the triangle; therefore, it is a midsegment.
Circumcenter	Three lengths	The distance from the point to the three vertices are equal, so it is equidistant from each vertex. Therefore, the point is the circumcenter.
Centroid	Two lengths	The distances from the point to the vertices and the distance from the point to the opposite midpoint are in a 1 : 2 ratio; therefore, the point is a centroid.

Practice Exercises

1 The coordinates of quadrilateral *BGRM* are $B(0, 2)$, $G(6, 0)$, $R(9, 9)$, and $M(3, 11)$.

   Prove *BGRM* is a rectangle.

2 $\triangle GYP$ has coordinates $G(1, 1)$, $Y(7, 5)$, and $P(3, 11)$.

   Prove $\triangle GYP$ is an isosceles right triangle.

3 Given parallelogram *FLIP* with vertices $F(2, 9)$, $L(h, k)$, $I(11, 6)$, and $P(1, 2)$, find the values of h and k.

4 Given quadrilateral *CUTE* with coordinates $C(-5, -4)$, $U(-3, 4)$, $T(5, 6)$, and $E(3, -2)$, prove *CUTE* is a rhombus.

5 Quadrilateral *PLOW* has coordinates $P(1, -4)$, $L(-1, 4)$, $O(7, 6)$, and $W(9, -2)$.

   Prove *PLOW* is a square.

6 Quadrilateral *ABCD* has coordinates $A(1, 2)$, $B(9, 4)$, $C(5, 8)$, and $D(1, 7)$.

   Prove *ABCD* is a trapezoid.

7 Segment $\overline{MG}$ has coordinates $M(0, 2)$ and $G(4, 0)$. Point T is the midpoint of $\overline{MG}$. $\overline{M''G''}$ is the image of $\overline{MG}$ after a 90° rotation about T followed by a dilation centered about T with a scale factor of 2.

a) State the coordinates of M'' and G'' and graph $\overline{M''G''}$.

b) Prove $MG''GM''$ is a rhombus.

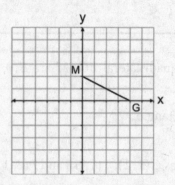

8 $\triangle ABC$ has coordinates $A(1, 5)$, $B(7, -1)$, and $C(13, 11)$. If point G has coordinates $(7, 5)$, prove G is the centroid of $\triangle ABC$.

9 $\triangle DOG$ has coordinates $D(4, 3)$, $O(9, 8)$, and $G(1, 12)$. If point A has coordinates $(3, 6)$, prove $\overline{OA}$ is an altitude of $\triangle DOG$.

10 Rectangle $ABCD$ has coordinates $A(4, 12)$, $B(13, 3)$, $C(9, -1)$, and $D(0, 8)$. Points F and G have coordinates $F(2, 10)$ and $G(7, 5)$, and $AHGF$ is also a rectangle. Is rectangle $ABCD$ similar to rectangle $AHGF$? Justify your answer.

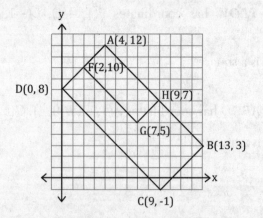

Solutions

1 There are several different approaches to proving a quadrilateral is a rectangle. Here, we will calculate the slopes of the 4 sides. This will let us show opposite sides have the same slope and are parallel, and consecutive sides have negative reciprocal slopes and are perpendicular. *BGRM* is graphed in the accompanying diagram.

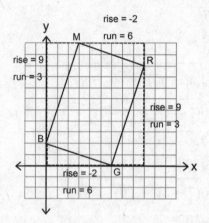

Using the grid, we can calculate slope from slope $= \dfrac{\text{rise}}{\text{run}}$. The rise and run for each side of *BGRM* are shown in the figure.

$$\text{Slope } \overline{BM} = \frac{\text{rise}}{\text{run}} \qquad \text{Slope } \overline{GR} = \frac{\text{rise}}{\text{run}}$$

$$= \frac{9}{3} \qquad\qquad = \frac{9}{3}$$

$$= 3 \qquad\qquad = 3$$

$$\text{Slope } \overline{BG} = \frac{\text{rise}}{\text{run}} \qquad \text{Slope } \overline{MR} = \frac{\text{rise}}{\text{run}}$$

$$= \frac{-2}{6} \qquad\qquad = \frac{-2}{6}$$

$$= -\frac{1}{3} \qquad\qquad = -\frac{1}{3}$$

The slopes of opposite sides are equal; therefore, they are parallel and *BGRM* is a parallelogram.

The slopes of consecutive sides are negative reciprocals, making them perpendicular. *BGRM* is a rectangle because it is a parallelogram with right angles.

2 The strategy is to show that $\overline{GY} \cong \overline{YP}$ using distance calculations and $\overline{GY} \perp \overline{YP}$ using slope calculations.

$$\text{slope} = \frac{y_2 - y_1}{x_2 - x_1}$$

$$\text{slope } \overline{GY} = \frac{5 - 1}{7 - 1} \qquad\qquad \text{slope } \overline{YP} = \frac{11 - 5}{3 - 7}$$

$$= \frac{4}{6} = \frac{2}{3} \qquad\qquad\qquad\qquad = -\frac{6}{4} = -\frac{3}{2}$$

$$\text{length} = \sqrt{\left(x_1 - x_2\right)^2 + \left(y_1 - y_2\right)^2}$$

$$\text{length } \overline{GY} = \sqrt{\left(7 - 1\right)^2 + \left(5 - 1\right)^2}$$

$$= \sqrt{36 + 16}$$

$$= \sqrt{52}$$

$$\text{length } \overline{YP} = \sqrt{\left(7 - 3\right)^2 + \left(5 - 11\right)^2}$$

$$= \sqrt{16 + 36}$$

$$= \sqrt{52}$$

$\overline{GY}$ is perpendicular to $\overline{YP}$ because their slopes are negative reciprocals, and $\overline{GY} \cong \overline{YP}$ because the lengths GY and YP are equal. Therefore, $\triangle GYP$ is a right isosceles triangle.

3. The opposite sides of a parallelogram are parallel and congruent. The strategy is to locate point L so that $FL = PI$ and the slopes of $\overline{FL}$ and $\overline{PI}$ are equal.

$$\text{slope} = \frac{y_2 - y_1}{x_2 - x_1}$$

$$\text{slope } \overline{PI} = \frac{6 - 2}{11 - 1}$$

$$= \frac{4}{10}$$

$$\text{slope } \overline{FL} = \frac{4}{10} \qquad \text{Parallel lines have equal slopes.}$$

Since slope is also equal to $\frac{\text{rise}}{\text{run}}$, locate point L by adding the rise to the y-coordinate of F, and add the run to the x-coordinate of x.

$$x\text{-coordinate of } L = 2 + 10$$
$$= 12$$
$$y\text{-coordinate of } L = 9 + 4$$
$$= 13$$

The coordinates of L are (12, 13).

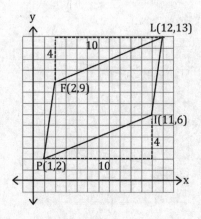

4 One strategy is to first prove the figure is a parallelogram by showing the diagonals bisect each other using midpoint calculations. Then prove it is a rhombus by showing the diagonals are perpendicular using slope calculations.

$$\text{midpoint} = \frac{x_1 + x_2}{2}, \frac{y_1 + y_2}{2}$$

midpoint $\overline{CT}$: $\frac{-5+5}{2}, \frac{-4+6}{2}$ midpoint $\overline{EU}$: $\frac{3+(-3)}{2}, \frac{-2+4}{2}$

$(0,1)$ $(0,1)$

$$\text{slope} = \frac{y_2 - y_1}{x_2 - x_1}$$

slope $\overline{CT} = \frac{6-(-4)}{5-(-5)}$ slope $\overline{EU} = \frac{4-(-2)}{-3-3}$

$= \frac{10}{10}$ $= -\frac{6}{6}$

$= 1$ $= -1$

The diagonals of *CUTE* bisect each because they have the same midpoint, making it a parallelogram. The diagonals are also perpendicular because their slopes are negative reciprocals. Therefore, *CUTE* is a rhombus.

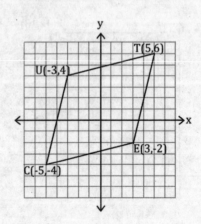

5 First prove the figure is a parallelogram by showing the diagonals bisect each other using midpoint calculations; then show it is a rhombus with perpendicular diagonals using slope. Finally, show it is a rectangle with a pair of perpendicular sides using slope.

$$\text{midpoint} = \frac{x_1 + x_2}{2}, \frac{y_1 + y_2}{2}$$

midpoint $\overline{PO}$: $\frac{1+7}{2}, \frac{-4+6}{2}$ midpoint $\overline{WL}$: $\frac{9+(-1)}{2}, \frac{-2+4}{2}$

$(4, 1)$ $(4, 1)$

$$\text{slope} = \frac{y_2 - y_1}{x_2 - x_1}$$

slope $\overline{PO} = \frac{6 - (-4)}{7 - 1}$ slope $\overline{WL} = \frac{4 - (-2)}{-1 - 9}$

$$= \frac{10}{6} \qquad\qquad\qquad = -\frac{6}{10}$$

slope $\overline{PW} = \frac{-2 - (-4)}{9 - 1}$ slope $\overline{LP} = \frac{-4 - 4}{1 - (-1)}$

$$= \frac{2}{8} \qquad\qquad\qquad = -\frac{8}{2}$$

The diagonals of *PLOW* bisect each other because they have the same midpoint, making it a parallelogram. The diagonals of *PLOW* are perpendicular because their slopes are negative reciprocals, making it a rhombus. *PLOW* has a pair of perpendicular consecutive sides because $\overline{PW}$ and $\overline{LP}$ have slopes that are negative reciprocals, making it a rectangle. *PLOW* is both a rectangle and a rhombus, so it is also a square.

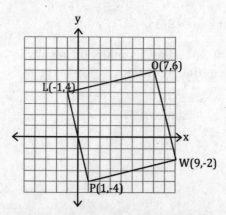

6 A good time-saving strategy is to graph the figure first to determine the pair of sides that are likely to be parallel. The parallel sides appear to be $\overline{AB}$ and $\overline{CD}$, and the legs are $\overline{AD}$ and $\overline{BC}$.

$$\text{slope} = \frac{y_2 - y_1}{x_2 - x_1}$$

$$\text{slope } \overline{AB} = \frac{4 - 2}{9 - 1} \qquad \text{slope } \overline{CD} = \frac{8 - 7}{5 - 1}$$

$$= \frac{2}{8} \qquad\qquad\qquad = \frac{1}{4}$$

$$= \frac{1}{4}$$

$\overline{AB}$ is parallel to $\overline{CD}$ because their slopes are equal, making $ABCD$ a trapezoid.

7 a) First find the coordinates of the midpoint T using the midpoint formula; then use T to graphically find the coordinates of M' and G'.

$$x_{MP} = \frac{1}{2}(x_1 + x_2)$$

$$= \frac{1}{2}(0 + 4)$$

$$= 2$$

$$y_{MP} = \frac{1}{2}(y_1 + y_2)$$

$$= \frac{1}{2}(2 + 0)$$

$$= 1$$

The coordinates of T are $(2, 1)$.

Rotating a segment 90° about its midpoint results in a segment congruent and perpendicular to the preimage.

$$\text{slope of } \overline{MG} = \frac{y_2 - y_1}{x_2 - x_1}$$

$$= \frac{0 - 2}{4 - 0}$$

$$= -\frac{1}{2}$$

Let $\overline{M'G'}$ be the image of $\overline{MG}$ after the 90° rotation. The slope of $\overline{M'G'}$ is the negative reciprocal, or 2. The coordinates of M' and G' can be found graphically by starting at T and moving with a slope of 2 in either direction for a length equal to TM and TG. From the graph, the coordinates are $M'(1, -1)$ and $G'(3, 3)$.

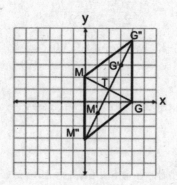

A dilation through midpoint T will extend the segment the same distance on either side of the midpoint because T remains the midpoint. Graph $\overline{M''G''}$ collinear with $\overline{M'G'}$, but twice as long.

b) $MG''GM''$ must be a rhombus. T is the midpoint of both of its diagonals. The diagonals bisect each other, making it a parallelogram. The 90° rotation resulted in $\overline{MG} \perp \overline{M''G''}$. A parallelogram with perpendicular diagonals is a rhombus.

8 The centroid is the point of concurrency of the medians of a triangle, so one strategy is to graph two medians and look for the point of intersection. A median is the segment from a vertex to the opposite midpoint, so the first step is to find the midpoints of two sides.

$$\text{midpoint} = \frac{x_1 + x_2}{2}, \frac{y_1 + y_2}{2}$$

let D be the midpoint of $\overline{AB}$: $\dfrac{1+7}{2}, \dfrac{5+(-1)}{2}$

$D(4,2)$

let E be the midpoint of $\overline{AC}$: $\dfrac{1+13}{2}, \dfrac{5+11}{2}$

$E(7,8)$

Graphing $\overline{CD}$ and $\overline{BE}$, the point of intersection is $G(7,5)$. Therefore, G is the centroid of $\triangle ABC$.

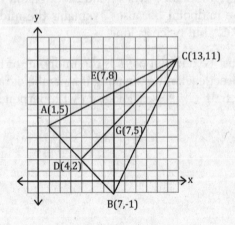

Algebraically, you could have also used the distance formula to calculate DG and GC and then show $DG = \frac{1}{2}GC$.

9 An altitude is a segment from a vertex perpendicular to the opposite side of a triangle. $\overline{AO}$ is an altitude if it is perpendicular to $\overline{GD}$.

$$\text{slope} = \frac{y_2 - y_1}{x_2 - x_1}$$

$$\text{slope } \overline{AO} = \frac{8 - 6}{9 - 3} \qquad \text{slope } \overline{GD} = \frac{12 - 3}{1 - 4}$$

$$= \frac{2}{6} \qquad\qquad\qquad = -\frac{9}{3}$$

$$= \frac{1}{3} \qquad\qquad\qquad = -\frac{3}{1}$$

$\overline{AO}$ is perpendicular to $\overline{GD}$ because their slopes are negative reciprocals, so $\overline{AO}$ is an altitude of $\triangle DOG$.

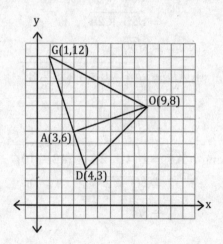

10 Since both figures are rectangles, all pairs of corresponding angles must be congruent. The lengths of two pairs of corresponding sides must be calculated to see if they are in the same ratio. The other two pairs must be in the same ratio since opposite sides of a rectangle are congruent.

The corresponding lengths to be checked are *AB* and *AH*, and *AD* and *AF*.

$$\text{length} = \sqrt{(x_1 - x_2)^2 + (y_1 - y_2)^2}$$

$$\text{length } AB = \sqrt{(4 - 13)^2 + (12 - 3)^2}$$

$$= \sqrt{81 + 81}$$

$$= \sqrt{162}$$

$$\text{length } AH = \sqrt{(4 - 9)^2 + (12 - 7)^2}$$

$$= \sqrt{25 + 25}$$

$$= \sqrt{50}$$

$$\text{length } AD = \sqrt{(4 - 0)^2 + (12 - 8)^2}$$

$$= \sqrt{16 + 16}$$

$$= \sqrt{32}$$

$$\text{length } AF = \sqrt{(4 - 2)^2 + (12 - 10)^2}$$

$$= \sqrt{4 + 4}$$

$$= \sqrt{8}$$

Calculate the ratios of the two pairs of corresponding sides.

$$\frac{AB}{AH} = \frac{\sqrt{162}}{\sqrt{50}} \qquad \frac{AD}{AF} = \frac{\sqrt{32}}{\sqrt{8}}$$

$$= \sqrt{\frac{81}{25}} \qquad\qquad = \sqrt{4}$$

$$= \frac{9}{5} \qquad\qquad\quad = 2$$

The rectangles are not similar because the corresponding sides are not in the same ratio.

3.11 CIRCLES

DEFINITIONS AND BASIC THEOREMS

Segment and Line Definitions in Circles

Segment or Line	Definition
Radius	A segment with one endpoint at the center of the circle and one endpoint on the circle
Chord	A segment with both endpoints on the circle
Diameter	A chord that passes through the center of the circle
Secant	A line that intersects a circle at exactly two points
Tangent	A line that intersects a circle at exactly one point
Point of tangency	The point at which a tangent intersects a circle

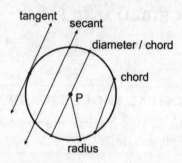

Some basic theorems of circles:

- All radii of a given circle are congruent.
- All circles are similar.
- Two circles are congruent if and only if their radii are congruent.

Since all circles are similar, any circle can be mapped to another using a similarity transformation comprised of a translation to move one center onto the other followed by a dilation to make the radii congruent.

An arc is a section of a circle. An arc may be major, minor, or a semicircle:

- *Minor arc*—an arc spanning less than 180°
- *Semicircle*—an arc spanning exactly 180°
- *Major arc*—an arc spanning more than 180°

When naming a semicircle or major arc, we add an additional point to show which arc is being specified. The accompanying figure shows a minor arc $\overset{\frown}{AB}$, semicircle $\overset{\frown}{ABC}$, and major arc $\overset{\frown}{ABD}$.

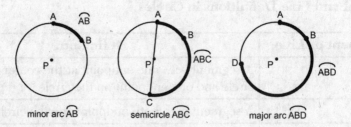

minor arc $\overset{\frown}{AB}$ semicircle $\overset{\frown}{ABC}$ major arc $\overset{\frown}{ABD}$

Some useful arc theorems:

- A diameter intercepts a semicircle.
- The sum of the arc measures around a circle equals 360°.
- The sum of the arc measures around a semicircle equals 180°.

CENTRAL AND INSCRIBED ANGLES

Central angle—an angle whose vertex is the center of a circle and whose rays intersect the circle

CENTRAL ANGLE THEOREM
The angle measure of an arc equals the measure of the central angle that intercepts the arc.

Inscribed angle—an angle whose vertex is on a circle and whose rays intersect the circle

INSCRIBED ANGLE THEOREM
The angle measure of an arc equals twice the measure of the inscribed angle that intercepts the arc.

The accompanying figure shows central angle $\angle APB$ and inscribed angle $\angle CMD$. Each angle cuts off, or intercepts, an arc on the circle. Central $\angle APB$ intercepts $\overarc{AB}$ and inscribed angle $\angle CMD$ intercepts $\overarc{CD}$.

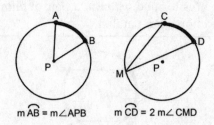

$m\,\overarc{AB} = m\angle APB$ $m\,\overarc{CD} = 2\,m\angle CMD$

- Since the inscribed angle measures half the intercepted arc, any inscribed angle that intercepts a semicircle is a right angle.

CONGRUENT, PARALLEL, AND PERPENDICULAR CHORDS

CONGRUENT CHORD THEOREM

Congruent chords intercept congruent arcs on a circle.
   Congruent arcs on a circle are intercepted by congruent chords.

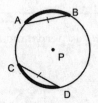

$$\overline{AB} \cong \overline{CD}, \quad \overarc{AB} \cong \overarc{CD}$$

PARALLEL CHORD THEOREM

The two arcs formed *between* a pair of parallel chords are congruent.

    If the two arcs formed *between* a pair of chords are congruent, then the chords are parallel.

$$\overline{AB} \parallel \overline{CD}, \quad \overset{\frown}{AC} \cong \overset{\frown}{BD}$$

CHORD–PERPENDICULAR BISECTOR THEOREM

The perpendicular bisector of any chord passes through the center of the circle.

    A diameter or radius that is perpendicular to a chord bisects the chord.

    A diameter or radius that bisects a chord is perpendicular to the chord.

In the accompanying figure, $\overline{AC}$ is perpendicular to diameter $\overline{BD}$ at E; therefore, E is the midpoint of $\overline{AC}$.

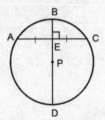

You can think of this theorem linking three properties of a diameter:

- Bisects another chord
- Is perpendicular to another chord
- Is a diameter

If any two are true, then the third must also be true.

TANGENT RADIUS THEOREM

- A diameter or radius to a point of tangency is perpendicular to the tangent.
- A line perpendicular to a tangent at the point of tangency passes through the center of the circle.

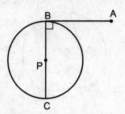

CONGRUENT TANGENT THEOREM

Given a circle and external point Q, segments between the external point and the two points of tangency are congruent.

In the accompanying figure, tangents $\overrightarrow{QA}$ and $\overrightarrow{QB}$ are both constructed from point Q, and $\overline{QA} \cong \overline{QB}$.

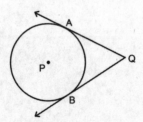

CYCLIC QUADRILATERALS

Quadrilaterals that are inscribed in a circle are called cyclic quadrilaterals.

> **CYCLIC QUADRILATERAL THEOREM**
>
> Opposite angles of cyclic quadrilaterals are supplementary.

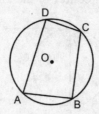

$\angle A$ and $\angle C$ are supplementary.
$\angle B$ and $\angle D$ are supplementary.

This property is proven by considering the intercepted arcs. Together, arcs $\overset{\frown}{ADC}$ and $\overset{\frown}{ABC}$ comprise the entire circle, so $m\overset{\frown}{ADC} + m\overset{\frown}{ABC} = 360°$. Since the intercepted angles measure $\frac{1}{2}$ the arc measures, $m\angle ABC + m\angle ADC = 180°$, and the angles are supplementary. The same justification can be used to show $\angle BAD$ and $\angle BCD$ are supplementary.

ANGLES FORMED BY INTERSECTING CHORDS, SECANTS, AND TANGENTS

Angles formed by intersecting chords, secants, and tangents follow three different relationships, depending on whether the vertex of the angle lies within, on, or outside the circle.

Vertex inside the circle	angle measure $= \frac{1}{2}$ the *sum* of the intercepted arcs $$m\angle 1 = m\angle 2 = \frac{1}{2}\left(m\widehat{AC} + m\widehat{DB}\right)$$ $$m\angle 3 = m\angle 4 = \frac{1}{2}\left(m\widehat{BC} + m\widehat{AD}\right)$$
Vertex on the circle 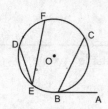	angle measure $= \frac{1}{2}$ of the intercepted arcs $$m\angle ABC = \frac{1}{2}m\widehat{BC}$$ $$m\angle DEF = \frac{1}{2}m\widehat{DF}$$
Vertex outside the circle 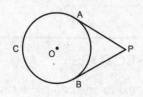 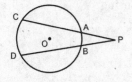 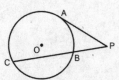	angle measure $= \frac{1}{2}$ the *difference* of the intercepted arcs $$m\angle P = \frac{1}{2}\left(m\widehat{ACB} - m\widehat{AB}\right)$$ $$m\angle P = \frac{1}{2}\left(m\widehat{CD} - m\widehat{AB}\right)$$ $$m\angle P = \frac{1}{2}\left(m\widehat{AC} - m\widehat{AB}\right)$$

SEGMENT RELATIONSHIPS IN INTERSECTING CHORDS, TANGENTS, AND SECANTS

The relationships between segment lengths formed by intersecting chords, tangents, and secants are shown in the following table.

Intersecting chords	products of the parts are equal $$a \cdot b = c \cdot d$$
Intersecting secants	outside · whole = outside · whole $$PA \cdot PC = PB \cdot PD$$
Intersecting tangent and secant	outside · whole = outside · whole $$PA^2 = PB \cdot PC$$

RADIAN MEASURE, ARC LENGTH, AND SECTORS

The **radian** is a unit of angle measure. 2π radians is equal to one complete revolution around a circle, or $360°$. Convert between units using:

$$\text{radians} = \frac{\pi}{180°} \cdot \text{degrees}$$

$$\text{degrees} = \frac{180°}{\pi} \cdot \text{radians}$$

- Besides having an angle measure, an arc has a length. The arc and its central angle also enclose a region called a sector.

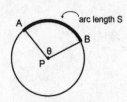

Arc with length S
intercepted by angle θ

Sector APB intercepted
by an angle θ

Arc length and sector area can be calculated from the measure of the central angle and the radius of the circle using the following formulas:

Unit of Angle Measure	Arc Length	Sector Area
Degrees	$\dfrac{\pi}{180°} R \cdot \theta$	$\dfrac{1}{360°} \pi R^2 \cdot \theta$
Radians	$R \cdot \theta$	$\dfrac{1}{2} R^2 \cdot \theta$

θ is the measure of the central angle.

Practice Exercises

1 Which of the following is a precise definition of a circle?
   (1) the set of points a fixed distance away from a given point
   (2) the set of points equidistant from the two endpoints of a given diameter
   (3) the set of points equidistant from a given center point and a given line
   (4) a closed figure with no angles

2 $\triangle STY$ is inscribed in circle P. Which of the following is *not necessarily* true?
   (1) Point P is equidistant from S, T, and Y.
   (2) The perpendicular bisector of $\overline{ST}$ passes through P.
   (3) The measure of angle Y equals one-half the measure of $\overset{\frown}{ST}$.
   (4) A tangent to the circle at S forms a right angle with $\overline{ST}$.

3 $\triangle FLY$ is inscribed in circle O. If $FL = LY$ and $m\overset{\frown}{FY} = 80°$, what is the measure of $\angle F$?
   (1) $10°$ (3) $70°$
   (2) $40°$ (4) $80°$

4 In $\odot P$, $\angle APB$ is a central angle and $\angle ADB$ is an inscribed angle. If $m\angle APB = 110°$, find $m\angle ADB$.

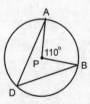

   (1) $55°$ (3) $70°$
   (2) $60°$ (4) $80°$

5 The clock in the tower of the Ridgefield town hall is in the shape of a circle 15 feet in diameter. What is the length of the arc around the clock whose central angle is defined by the hour and minute hands at 4:00 P.M.?

(1) 15.7 ft

(2) 18.6 ft

(3) 31.4 ft

(4) 60 ft

6 In circle Z, arc $\overset{\frown}{KP}$ is intercepted by central angle $\angle KZP$. If m$\angle KZP$ = 1.2 radians and $ZK = 7$ cm, what is the length of $\overset{\frown}{KP}$?

(1) 5.8 cm

(2) 8.4 cm

(3) 10.1 cm

(4) 26.4 cm

7 Circle O has a radius of 14 inches. If $\angle HOT$ is a central angle in circle O that measures 132°, what is the area of sector HOT?

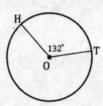

(1) 4.7 in^2

(2) 10 in^2

(3) 226 in^2

(4) 294 in^2

8 Given ⊙P with radius $\overline{PB}$, tangent $\overline{AB}$, and m∠$P = 53°$, what is the measure of ∠A?

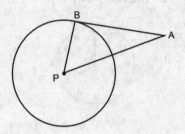

(1) 26.5°   (3) 53°
(2) 37°   (4) 60°

9 In ⊙P, chords $\overline{AB}$ and $\overline{CD}$ are parallel. If m$\overset{\frown}{AB} = 118°$ and m$\overset{\frown}{CD} = 78°$, find m$\overset{\frown}{AC}$.

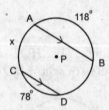

10 In ⊙P, chords $\overline{AB}$ and $\overline{CD}$ are congruent, m$\overset{\frown}{BC} = 60°$, and m$\overset{\frown}{CD} = 80°$. Find m∠$ABD$.

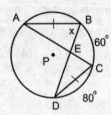

11 $\overline{BDP}$ is a radius of $\odot P$ and is perpendicular to $\overline{AC}$. If $AC = 24$ and $PC = 13$, find PD and DB.

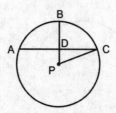

12 $\triangle RST$ is circumscribed about circle P. If $SX = 6$, $XR = 7$, and $TZ = 8$, what is the perimeter of $\triangle RST$?

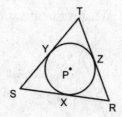

13 Secant $\overline{BEA}$ intersects circle O at E and A. Secant $\overline{BDR}$ intersects circle O at points D and R. If $m\widehat{AR} = 78°$ and $m\widehat{ED} = 35°$, what is the measure of $\angle B$?

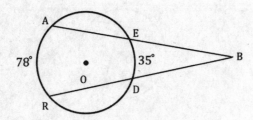

14 Secant $\overline{FGH}$ intersects circle O at G and H. Secant $\overline{FPQ}$ intersects circle O at P and Q. If $FG = 6$, $GH = 10$, and $FP = 8$, find the length of PQ.

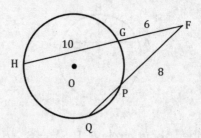

15 Tangents $\overline{TR}$ and $\overline{TQ}$ are drawn from point T to circle A. $\overline{RAS}$ is a diameter of circle A and $m\angle T = 46°$. What is the measure of $\overset{\frown}{SQ}$?

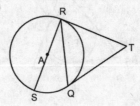

16 In $\odot O$, chords $\overline{HLI}$ and $\overline{KLJ}$ intersect at L. Prove: $HL \cdot LI = KL \cdot LJ$

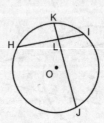

Solutions

1 Examine each choice:

(1) Correct—The fixed distance is the radius of the circle, and the given point is the center of the circle.

(2) Incorrect—The set of points equidistant from the two endpoints of a diameter would be the perpendicular bisector of the diameter.

(3) Incorrect—The set of points equidistant from a center point and line would be a parabola.

(4) Incorrect—Circles are closed figures with no angles, but not all closed figures with no angles are circles. A counterexample would be an ellipse.

The correct choice is **(1)**.

2 Choice (1) is true—The center of the circumscribed circle is equidistant from the vertices of the triangle.

Choice (2) is true—A perpendicular bisector of a chord always passes through the center of the circle.

Choice (3) is true—The measure of the inscribed angle equals half the measure of the intercepted arc.

Choice (4) is not necessarily true—A tangent will make a right angle with a chord only if the chord is a diameter. $\overline{ST}$ is not necessarily a diameter.

The correct choice is **(4)**.

3 Sketching the isosceles triangle and labeling the known arc, we see $\angle L$ is the inscribed angle that intercepts the 80° arc.

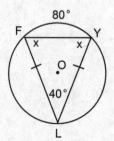

An inscribed angle measures half the intercepted arc:

$$m\angle L = \frac{1}{2}m\widehat{FY}$$

$$= \frac{1}{2}\left(80°\right)$$

$$= 40°$$

Since $FL = LY$, $m\angle F = m\angle Y$ from the isosceles triangle theorem, and we can use the angle sum theorem to find the measures.

$$m\angle F + m\angle L + m\angle Y = 180°$$

$$2x + 40° = 180°$$

$$2x = 140°$$

$$x = 70°$$

$$m\angle F = 70°$$

The correct choice is **(3)**.

4 The strategy is to use the central angle to find $m\widehat{AB}$, and then use the arc to find the inscribed angle.

$m\widehat{AB} = m\angle APB = 110°$ The central angle and arc have equal measure.

$m\angle ADB = \frac{1}{2}m\widehat{AB} = 55°$ The inscribed angle measure is half the intercepted arc.

The correct choice is **(1)**.

5 The numbers on the face of the clock divide the circle into 12 congruent central angles. Each central angle measures $\dfrac{360°}{12}$, or 30°. Therefore, the angle formed by the two hands on the clock measures $4 \cdot 30°$, or 120°. The radius $= \dfrac{1}{2}$ (15), or 7.5 ft. Using the formula for arc length,

$$S = 2\pi R \dfrac{\theta}{360°}$$

$$= 2\pi \left(7.5\,\text{ft}\right)\dfrac{120°}{360°}$$

$$= 15.7\,\text{ft}$$

The correct choice is (1).

6 Arc length, S, is given by the equation $S = R\theta$, where θ is the measure of the central angle in radians.

$$S = R\theta$$

$$= 7\,\text{cm} \cdot 1.2$$

$$= 8.4\,\text{cm}$$

The correct choice is (2).

7 The area of a sector is given by $A = \dfrac{\theta}{360°}\pi R^2$, where θ is the central angle measured in degrees.

$$A = \dfrac{132°}{360°}\pi \cdot \left(14\,\text{in}\right)^2$$

$$= 225.775\,\text{in}^2$$

$$= 226\,\text{in}^2$$

The correct choice is (3).

8                    $m\angle B = 90°$ radius perpendicular to a tangent
                                              at the point of tangency

$m\angle P + m\angle B + m\angle A = 180°$ triangle angle sum theorem

$53° + 90° + m\angle A = 180°$

$143° + m\angle A = 180°$

$m\angle A = 37°$

The correct choice is **(2)**.

9 Arcs $\overset{\frown}{AC}$ and $\overset{\frown}{DB}$ are the congruent arcs between a pair of parallel chords.

$m\overset{\frown}{AB} + m\overset{\frown}{BD} + m\overset{\frown}{CD} + m\overset{\frown}{AC} = 360°$ circle–arc sum theorem

$118° + x + 78° + x = 360°$ parallel chord theorem

$196° + 2x = 360°$

$2x = 164°$

$x = 82°$

$m\overset{\frown}{AC} = 82°$

10 $\angle ABD$ is an inscribed angle, so its measure will be half of $m\overset{\frown}{AD}$.

$m\overset{\frown}{AB} + m\overset{\frown}{BC} + m\overset{\frown}{CD} + m\overset{\frown}{AD} = 360°$ circle–arc sum theorem

$m\overset{\frown}{AB} = m\overset{\frown}{CD} = 80°$ congruent chord theorem

$80° + 60° + 80° + m\overset{\frown}{AD} = 360°$

$220° + m\overset{\frown}{AD} = 360°$

$m\overset{\frown}{AD} = 140°$

$m\angle ABD = \frac{1}{2}m\overset{\frown}{AD}$

$= 70°$

11 Radius $\overline{BDP}$ is perpendicular to $\overline{AC}$, so it must also bisect $\overline{AC}$. The strategy is to calculate DC and then apply the Pythagorean theorem in $\triangle PDC$.

$$DC = \frac{1}{2}AC$$

$$DC = \frac{1}{2}(24)$$

$$DC = 12$$

$$(PD)^2 + (DC)^2 = PC^2 \qquad \text{Pythagorean theorem}$$

$$(PD)^2 + 12^2 = 13^2$$

$$(PD)^2 + 144 = 169$$

$$(PD)^2 = 25$$

$$PD = 5$$

$$PB = PC = 13 \quad \text{all radii of a circle are congruent}$$

$$PD + DB = PB \qquad \text{partition}$$

$$5 + DB = 13$$

$$DB = 8$$

12 $SY = SX$, $TY = TZ$, and $ZR = XR$ because each pair are tangents from the same point.

The perimeter is, therefore, $2 \cdot 6 + 2 \cdot 7 + 2 \cdot 8 = 42$.

13 The angle formed by two secants is equal to half the difference of the intercepted arcs.

$$m\angle B = \frac{1}{2}\left(m\widehat{AB} - m\widehat{ED}\right)$$

$$= \frac{1}{2}\left(78° - 35°\right)$$

$$= 21.5°$$

14 When two tangents from the same point intersect a circle, the segments formed follow the relationship:

$$\text{outside} \cdot \text{whole} = \text{outside} \cdot \text{whole}$$
$$FG \cdot FH = FP \cdot FQ$$
$$6 \cdot (6 + 10) = 8 \cdot (8 + PQ)$$
$$6 \cdot 16 = 8(8 + PQ)$$
$$96 = 64 + 8PQ$$
$$32 = 8PQ$$
$$PQ = 4$$

15 This problem requires applying multiple circle concepts. The strategy is to first use the fact that $\triangle TRQ$ is isosceles to find $m\angle TRQ$ and $m\angle SRQ$ and then to apply the relationship that the measure of $\overset{\frown}{SQ}$ is twice the inscribed angle that intercepts it.

$\overline{TR} \cong \overline{TQ}$	congruent tangent theorem
$m\angle TRQ = \frac{1}{2}\left(180° - m\angle T\right)$	isosceles triangle theorem in $\triangle TRQ$
$m\angle TRQ = \frac{1}{2}\left(180° - 46°\right)$	
$= 67°$	

$m\angle SRT = 90°$	tangent $RT \perp$ diameter $\overline{RAS}$
$m\angle SRQ + m\angle TRQ = 90°$	$\angle SRQ$ and $\angle TRQ$ are complementary
$m\angle SRQ + 67° = 90°$	
$m\angle SRQ = 23°$	
$m\overset{\frown}{SQ} = 2m\angle SRQ$	inscribed angle theorem
$m\overset{\frown}{SQ} = 2\left(23°\right)$	
$m\overset{\frown}{SQ} = 46°$	

16 The strategy is to sketch $\overline{HJ}$ and $\overline{KI}$, show $\triangle HLJ \sim \triangle KLI$, and then write a proportion between corresponding sides.

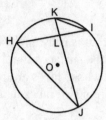

Statement	Reason
1. Chords $\overline{HLI}$ and $\overline{KLJ}$ intersect at L	1. Given
2. Construct $\overline{IK}$ and $\overline{HJ}$	2. Two points define a segment
3. $\angle H \cong \angle K$ $\angle J \cong \angle I$	3. Inscribed angles that intercept the same arc are congruent
4. $\triangle HLJ \sim \triangle KLI$	4. AA
5. $\dfrac{HL}{KL} = \dfrac{LJ}{LI}$	5. Corresponding parts of similar triangles are proportional
6. $HL \cdot LI = KL \cdot LJ$	6. Cross products of a proportion are equal

3.12 SOLIDS

DEFINITIONS

- A *solid* is any 3-D figure that is fully enclosed.
- A *face* of a solid is any of the surfaces that bound the solid.
- A *polyhedron* is any solid whose faces are polygons.
- An *edge* is the intersection of two faces in a polyhedron.

PRISMS

A *prism* is a polyhedron with two congruent, parallel polygons for bases. The bases of a prism can have any shape. The accompanying figure shows 6 different prisms.

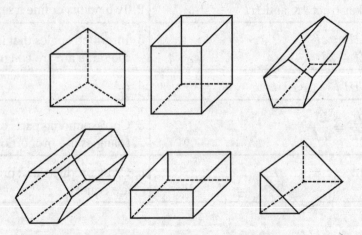

When working with prisms, keep in mind the following facts:

- The height of a prism, h, is the distance between the two bases shown in the accompanying figure.
- The lateral faces are all the faces other than the two parallel bases.
- A right prism has lateral edges that are perpendicular to the bases and lateral faces that are rectangles.
- The lateral edges of an oblique prism are not perpendicular to the bases, and the lateral faces are parallelograms.
- The volume of any prism is found with the formula $V = Bh$, where B is the area of the base.

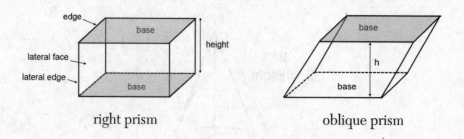

right prism   oblique prism

CYLINDERS

A *circular cylinder* is a solid figure with two parallel and congruent circular bases and a curved lateral area. The height is the perpendicular distance between the bases. As with the prisms, cylinders can be right or oblique as shown in the accompanying figure.

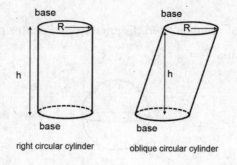

right circular cylinder   oblique circular cylinder

The volume of a cylinder is given by $V = Bh$, where B is the area of the base. In a circular cylinder, the circular base has an area of πR^2, so the volume formula can be rewritten as $V = \pi R^2 h$.

CONES AND PYRAMIDS

A *circular cone* is a solid with one circular base that comes to a point at an apex. A *pyramid* is a polyhedron having one polygonal base and triangles for lateral faces. The base of a pyramid can be any polygon, and the lateral faces are all triangles. The height of cones and pyramids is the perpendicular distance from the apex to the base. The slant height is the distance along the lateral surface perpendicular to the perimeter of the base.

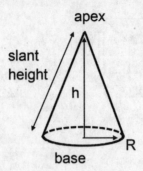

The volume of pyramids and cones are found from the same formula:

$V = \frac{1}{3}Bh$ where B is the area of the base.

SPHERES

A *sphere* is the set of points a fixed distance from a center point, and its volume is given by the formula $V = \frac{4}{3}\pi R^3$.

SURFACE AREA AND LATERAL AREA

The *surface area* is the area of all faces of a solid. Surface area can be found by calculating the area of all the faces of a solid individually and then summing them. You should be able to calculate the surface area of cubes, prisms, and pyramids since the faces of these solids are all polygons. Surface area of cones, cylinders, and spheres involve curved surfaces and are outside the scope of this course.

The lateral area of a solid excludes the bases. For a prism, do not include the two parallel bases. For a pyramid, exclude the one base.

Lateral face—any face of a solid other than its bases
Lateral area—the area of all the lateral faces of a solid
Surface area—the total area of a solid, lateral area + area of bases

CROSS-SECTIONS

A cross-section of a solid is the two-dimensional figure created when a plane intercepts a solid. Cross-sections are often taken parallel or perpendicular to the base of a figure. The shape of the cross-section depends on the angle at which the plane intersects the solid.

Cross-sections of a pyramid, sphere, and cylinder are shown in the accompanying figure.

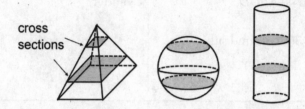

A cross-section of a pyramid taken perpendicular to the base will be shaped triangular, as shown in the figure, or trapezoidal. The cross-section of a cylinder perpendicular to the base is a rectangle.

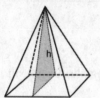 

SOLIDS OF REVOLUTION

Some solids can be generated by rotating a planar figure around a line. These are called *solids of revolution*. Some examples of solids of revolution and the figures that generate them are given in the following table.

Figure Rotated	Solid
Right triangle, rotate 360° about leg $\overline{AB}$	Cone
Rectangle $ABCD$, rotate 360° around side $\overline{CD}$	Cylinder
Circle P, rotate 180° about diameter $\overline{AB}$	Sphere

MODELING WITH SOLIDS

Many physical objects can be modeled using the basic solids discussed in this chapter. Some examples are:

Cylinders—trees, cans, barrels, people
Spheres—balls, balloons, planets
Prisms—bricks, boxes, aquariums, swimming pools, rooms, books

Volumes or areas can also be used to calculate some other quantity. Some common relationships to look for are

- Mass = density × volume
- Total cost = cost per unit volume × volume

   or

   Total cost = cost per unit area × area
- Energy contained in a material = heat content × volume
- Total population = population density × area

For example, if a can has a volume of 100 cm^3 and is filled with a liquid whose density is 1.2 grams/cm^3, then the mass of the liquid inside the can is

$$\text{Mass} = \text{density} \cdot \text{volume}$$
$$\text{Mass} = 1.2 \text{ grams/cm}^3 \cdot 100 \text{ cm}^3$$
$$= 120 \text{ grams}$$

In geometry design problems, we are asked to find the dimensions of some figure or solid that lets the object meet some given condition. Conditions that may be specified include

- maximum or minimum size
- a specified size
- maximum or minimum cost

Approach these problems in the same way as any other modeling problem. The difference comes in using the model to find a maximum or minimum. For example, you may have an equation that gives volume in terms of a length. A maximum or minimum can be found by

- *Trial and error/table*—Using a best first guess for the variable, evaluate the quantity. Change the value of the variable and evaluate the quantity again— it will increase or decrease. Continue until you see the quantity change from decreasing to increasing (a minimum) or increasing to decreasing (a maximum). The table feature of the graphing calculator is excellent for calculating many trials quickly. The table increment can be set to the desired precision.
- *Graphing*—The equation that described the quantity can be graphed, and the maximum/minimum feature of the calculator can then be applied to the graph.
- *Axis of symmetry*—If the equation that describes the quantity is *quadratic* in the form $y = ax^2 + bx + c$, then the maximum or minimum will always occur at the axis of symmetry $x = -b/2a$.

Practice Exercises

1 A basketball has a diameter of 9.6 inches. What is its volume?

   (1) 72.3 in^3 (3) 463 in^3

   (2) 347 in^3 (4) 3706 in^3

2 A right circular cylinder has a volume of 200 cm^3. If the height of the cylinder is 6 cm, what is the radius of the cylinder? Round your answer to the nearest hundredth.

   (1) 2.9 cm   (3) 3.3 cm

   (2) 3.0 cm   (4) 3.8 cm

3 Jenna is preparing to plant a new lawn and has a pile of topsoil delivered to her house. A dump truck delivers the topsoil and dumps it in a pile that is cone shaped. The radius of the pile is 9 ft, and the height is 3 ft. If the topsoil has a density of 95 lb/ft^3, what is the weight of the topsoil?

   (1) 18,764 lb   (3) 28,546 lb

   (2) 24,175 lb   (4) 32,676 lb

4 A right circular cylinder has a radius of 4 and a height of 10. What is the area of a cross-section taken parallel to, and halfway between, the bases? Write your answer in terms of π.

   (1) 4π (3) 12π

   (2) 8π (4) 16π

5 A cylinder and a cone have the same volume. If each solid has the same height, what is the ratio of the radius of the cylinder to the radius of the cone?

(1) $2:1$ (3) $\sqrt{2}:2$

(2) $3:1$ (4) $\sqrt{3}:3$

6 $\triangle JKL$ is a right triangle with a right angle at K, $JK = 6$, and $KL = 10$. Which of the following solids is generated with $\triangle JKL$ rotated 360° about KL?

(1) a cylinder with a height of 10 and a diameter of 12

(2) a cylinder with a height of 6 and a diameter of 20

(3) a right circular cone with a height of 10 and a diameter of 12

(4) a right circular cone with a height of 6 and a diameter of 20

7 The base of a cone is described by the curve $x^2 + y^2 = 16$. If the altitude of the cone intersects the center of its base, and the volume of the cone is 72π, what is the height of the cone?

8 A cylindrical piece of metal has a radius of 15 in and a height of 2 in. It goes through a hot-press machine that reduces the height of the cylinder to $\frac{1}{4}$ in. What is the new radius of the cylinder, assuming no material is lost?

9 Jack is making a scaled drawing of the floor plan of his home. The scale factor is 1 in : 4 ft. The drawing of his living room is a rectangle measuring 5 inches by 3 inches. He is planning to purchase new carpet for the living room that costs $4 per square foot. How much will the carpet cost?

(1) $720   (3) $960

(2) $840   (4) $1,040

10 The city of Deersfield sits along the bank of the Fox Run River. A map of the city is shown below. The downtown region, indicated by region A, is $\frac{1}{2}$ mile wide and 1 mile long. The population density of Deersfield, except for the downtown region, is 800 people per square mile. The population density of the downtown region is 4,000 people per square mile. What is the population of Deersfield? Round to the nearest whole number.

11 An engineer for a construction company needs to calculate the total volume of a home that is under construction so the proper sized heating and air conditioning equipment can be installed. The house is modeled as a rectangular prism with a triangular prism for the roof as shown in the accompanying figure. The roof is symmetric with $EF = ED$. The dimensions are $AB = 30$ ft, $BC = 50$ ft, $BD = 25$ ft, and the measure of $\angle EDF = 40°$. What is the total volume of the house? Round to the nearest cubic foot.

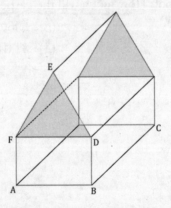

12 The Red Ribbon Orchard sorts their apple harvest by size. The three size categories are shown in the accompanying table.

	Grade B	Grade A	Grade AA
Average diameter	3 inches	4 inches	5 inches

After one day's harvest, there were 3,000 pounds of grade B apples, 4,200 pounds of grade A apples, and 3,200 pounds of grade AA apples. If the average density of an apple is 0.016 lb/in^3, what is the total number of apples that were harvested?

13 A construction company is working on plans for a project that call for digging a straight 0.25 mile tunnel through a mountain. The cross-section of the tunnel is shown in the accompanying figure.

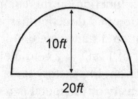

10*ft*

20*ft*

As the workers dig through the mountain, the rubble is brought to a gravel company to be processed into gravel. The cargo beds of the dump trucks are 18 ft long by 10 ft wide by 6 ft high, and the trucking company charges $450 per load delivered to the gravel company. Approximately how much money should the engineer budget for trucking costs for the project?

14 An engineer is designing a hinge to be used on the landing gear of a new airplane. She sketches the cross-section of the hinge on a computer, which is shown in the figure below. The circle represents a hole for a hinge pin.

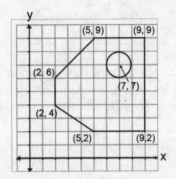

The engineer next uses a function on the computer that creates a solid by translating the cross-section in a direction perpendicular to the sketch. The engineer enters a translation distance of 10 units. In the final step, the engineer enters a scale factor of 1 unit = 2 centimeters. The resulting computer image of the solid is then sent to a factory to be manufactured out of a high-strength metal. If the density of the metal is 8.44 grams/cm^3 and the metal costs $32 per kilogram, what is the expected cost for the metal to make one hinge? Round to the nearest dollar.

Solutions

1 The basketball can be modeled as a sphere whose volume is $\frac{4}{3}\pi R^3$. The radius equals half the diameter, or 4.8 in.

$$V = \frac{4}{3}\pi R^3$$
$$= \frac{4}{3}\pi (4.8 \text{ in})^3$$
$$= 463 \text{ in}^3$$

The correct choice is **(3)**.

2 Start with the formula for volume, substitute the known values, and then solve for the radius.

$$V = B \cdot h$$
$$V = \pi R^2 h$$
$$200 = \pi \left(R^2\right)(6)$$
$$R^2 = 10.61032$$
$$R = \sqrt{10.61032}$$
$$= 3.3 \text{ cm}$$

The correct choice is **(3)**.

3 To find the weight of the topsoil, we need to first find the volume of the pile. The volume of a cone is $\frac{1}{3}Bh$, where B is the area of the base. This requires finding the area of the circular base.

$$B = \pi R^2$$
$$= \pi \left(9^2\right)$$
$$= 254.4690 \text{ ft}^2$$

We are now ready to find the volume.

$$V = \frac{1}{3}B \cdot h$$

$$= \frac{1}{3}(254.4690)(3)$$

$$= 254.469 \text{ ft}^3$$

Now calculate the weight using the volume and density.

$$\text{weight} = V \cdot \text{density}$$

$$= 254.469 \text{ ft}^3 \cdot 95 \text{ lb/ft}^3$$

$$= 24{,}174.55 \text{ lb}$$

$$= 24{,}175 \text{ lb}$$

The correct choice is (**2**).

4 The cross-section will be congruent to the base, so it is a circle with radius of 4. The area is πR^2, or 16π.

The correct choice is (**4**).

5 The strategy is to set the two volume formulas equal to each other and rearrange, solving for the ratio of the radii.

$$V_{\text{cone}} = \frac{1}{3}\pi R_{\text{cone}}^2$$

$$V_{\text{cylinder}} = \pi R_{\text{cylinder}}^2 h$$

$$V_{\text{cylinder}} = V_{\text{cone}}$$

$$\pi R_{\text{cylinder}}^2 h = \frac{1}{3}\pi R_{\text{cone}}^2 h$$

Now solve for the ratio $\dfrac{R_{\text{cylinder}}}{R_{\text{cone}}}$

$$R_{\text{cylinder}}^2 = \frac{1}{3} R_{\text{cone}}^2 \quad \text{divide by } \pi h$$

$$\frac{R_{\text{cylinder}}^2}{R_{\text{cone}}^2} = \frac{1}{3} \quad \text{divide by } R_{\text{cone}}^2$$

$$\frac{R_{\text{cylinder}}}{R_{\text{cone}}} = \frac{1}{\sqrt{3}} \quad \text{take square root of each side}$$

The ratio does not match any of the choices, so try rationalizing the denominator. To do so, multiply the numerator and denominator by $\sqrt{3}$.

$$\frac{R_{\text{cylinder}}}{R_{\text{cone}}} = \frac{1}{\sqrt{3}} \cdot \frac{\sqrt{3}}{\sqrt{3}}$$

$$\frac{R_{\text{cylinder}}}{R_{\text{cone}}} = \frac{\sqrt{3}}{3}$$

The correct choice is (**4**).

6 A right triangle rotated 360° about one of its legs will generate a cone.

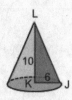

The leg aligned with the axis of rotation becomes the height, so the height equals 10. The leg perpendicular to the axis of rotation becomes the radius, so the radius equals 6 and the diameter equals 12.

The correct choice is (**3**).

7 $x^2 + y^2 = 16$ describes a circle centered at the origin with a radius of 4. Its area is

$$A = \pi R^2 = 16\pi$$

Apply the volume formula of the cone to find the height.

$$V = \frac{1}{3}Bh$$

$$72\pi = \frac{1}{3}16\pi h$$

$$h = 72 \cdot 3 \cdot \frac{1}{16}$$

$$= 13.5$$

8 Find the volume of the cylinder before pressing and set it equal to the expression for the volume after pressing. Use this equation to solve for the radius after pressing.

$$V = \pi R^2 h$$

$$V_{before} = \pi \cdot 15^2 \cdot 2$$

$$= 450\pi \ in^3$$

$$V_{after} = \pi \cdot R^2 \cdot \frac{1}{4} \qquad \text{we don't know the new radius}$$

$$V_{before} = V_{after}$$

$$\pi \cdot R^2 \cdot \frac{1}{4} = 450\pi \qquad \text{volumes must be equal}$$

$$\frac{R^2}{4} = 450$$

$$R^2 = 1{,}800$$

$$R = \sqrt{1{,}800}$$

$$= 42.426 \ in$$

9 The actual length and width of the living room are found by applying the scale factor to the drawing dimensions.

$$5 \text{ inches} \cdot \frac{4 \text{ ft}}{1 \text{ in}} = 20 \text{ ft}$$

$$3 \text{ inches} \cdot \frac{4 \text{ ft}}{1 \text{ in}} = 12 \text{ ft}$$

The area of the rectangular living room is

$$A = \text{length} \cdot \text{width}$$
$$= 20 \text{ ft} \cdot 12 \text{ ft}$$
$$= 240 \text{ ft}^2$$

The cost of the carpet is

$$\text{cost} = \text{area} \cdot \text{cost per ft}^2$$
$$\text{cost} = 240 \text{ ft}^2 \cdot \frac{\$4}{\text{ft}^2}$$
$$= \$960$$

The correct choice is **(3)**.

10 We can model the town as a semicircle and the downtown region as a rectangle. To calculate the population, we first find the area of each region. Let the area of the downtown region be represented by A_A and the area of the remainder of the town be represented by A_B.

$$A_\text{A} = \text{length} \cdot \text{width area of circle}$$
$$= \frac{1}{2} \cdot 1$$
$$= 0.5 \text{ mi}^2$$

The area of the remainder of the town is a semicircle minus the area of the downtown region.

$$A_\text{B} = \frac{1}{2}\pi R^2 - 0.5 \qquad \text{area of semicircle circle} - \text{area of rectangle}$$
$$= \frac{1}{2}\pi (2)^2 - 0.5$$
$$= 5.78318 \text{ mi}^2$$

The population of each region is the product of the area and the population density

$$\text{Population} = A_A \cdot 800 \text{ people/mi}^2 + A_B \cdot 4{,}000 \text{ people/mi}^2$$
$$= 5.78318 \text{ mi}^2 \cdot 800 \text{ people/mi}^2 + 0.5 \text{ mi}^2 \cdot 4{,}000 \text{ people/mi}^2$$
$$= 6{,}627 \text{ people}$$

The population of Deerfield is 6,627 people.

11 Calculate the volume of each section.

Rectangular prism:

$$\text{length} = BC \quad \text{width} = AB \quad \text{height} = BD$$
$$\text{length} = 50 \text{ ft} \quad \text{width} = 30 \text{ ft} \quad \text{height} = 25 \text{ ft}$$

$$\text{Volume}_{\text{rectangular prism}} = \text{length} \cdot \text{width} \cdot \text{height}$$
$$= (50 \text{ ft})(30 \text{ ft})(25 \text{ ft})$$
$$= 37{,}500 \text{ ft}^3$$

Triangular prism:

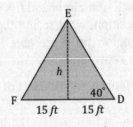

The first step is to find the area of the triangular base, $\triangle FED$. FD has a length equal to AB, or 30 ft. The height of the isosceles triangle divides the base into two 15 ft lengths. The tangent ratio can be used to find the height of $\triangle FED$.

$$\tan = \frac{\text{opposite}}{\text{adjacent}}$$

$$\tan(40°) = \frac{h}{15}$$

$$h = 15\tan(40°)$$

$$= 12.586 \text{ ft}$$

Now multiply the formula for the area of a triangle.

$$A_{\triangle FED} = 2 \cdot \frac{1}{2}\text{base} \cdot \text{height}$$

$$= 15 \text{ ft} \cdot 12.586 \text{ ft}$$

$$= 188.79 \text{ ft}^2$$

The volume of a triangular prism is equal to Bh, where B is the area of the triangular base and h is the length of the prism, which is equal to BC.

$$V_{\text{triangular prism}} = Bh$$

$$= A_{\triangle FED} \cdot BC$$

$$= 188.79 \text{ ft}^2 \cdot 50 \text{ ft}$$

$$= 9,439.5 \text{ ft}^3$$

Sum the two volumes to get the total volume of the house.

$$\text{Volume} = 37,500 \text{ ft}^3 + 9,439 \text{ ft}^3$$

$$= 46,939.5 \text{ ft}^3$$

$$= 46,940 \text{ ft}^3$$

12 The number of apples of each type equals the total volume of that type divided by the volume of one apple. Model the apples as spheres to calculate the volume of one apple. Find the total volume by dividing the weight by the density. The following table summarizes the calculations.

	Grade B	Grade A	Grade AA
volume of 1 apple = $V = \frac{4}{3}\pi R^3$	$= \frac{4}{3}\pi \cdot (1.5)^3$ $= 14.1372 \text{ in}^3$	$= \frac{4}{3}\pi \cdot (2)^3$ $= 33.5103 \text{ in}^3$	$= \frac{4}{3}\pi \cdot (2.5)^3$ $= 65.4498 \text{ in}^3$
Total volume of each type = $\dfrac{\text{weight}}{\text{density}}$	$= \dfrac{3,000 \text{ lb}}{0.016 \text{ lb/in}^3}$ $= 187,500 \text{ in}^3$	$= \dfrac{4,200 \text{ lb}}{0.016 \text{ lb/in}^3}$ $= 262,500 \text{ in}^3$	$= \dfrac{3,000 \text{ lb}}{0.016 \text{ lb/in}^3}$ $= 200,000 \text{ in}^3$
number of each type = $\dfrac{V_{\text{total}}}{V_{\text{one apple}}}$	$= \dfrac{187,500}{14.1372}$ $= 13,263$	$= \dfrac{262,500}{33.5013}$ $= 7,833$	$= \dfrac{200,000}{65.4498}$ $= 3,056$

The total number of apples is $13,263 + 7,833 + 3,056 = 24,152$ apples.

13 First, find the volume of the tunnel, and divide by the volume of the dump truck bed to find the number of dump-truck loads needed. The cross-section of the tunnel is a semicircle, which makes the tunnel in the shape of a half cylinder. The cylinder radius is 10 ft and the length is 0.25 miles.

First, convert 0.25 miles to feet using the conversion factor from the formula sheet.

$$0.25 \text{ mi} \cdot \frac{5,280 \text{ ft}}{\text{mi}} = 1,320 \text{ ft}$$

$$V_{\text{tunnel}} = \frac{1}{2}\pi R^2 h$$

$$= \frac{1}{2}\pi \cdot 10^2 \cdot 1,320$$

$$= 207,345.11 \text{ ft}^3$$

Model the dump-truck bed as a rectangular prism.

$$V_{truck} = \text{length} \cdot \text{width} \cdot \text{height}$$
$$= 18 \cdot 10 \cdot 6 = 1,080 \text{ ft}^3$$

The number of dump-truck loads is the ratio of the two volumes.

Number of loads $= V_{tunnel}/V_{truck}$

$$= \frac{207,345.11}{1,080}$$

$$= 191.98 \text{ dump-truck loads}$$

192 dump-truck loads are needed.

The estimated cost $= \dfrac{\$450}{\text{load}} \cdot 192 \text{ loads} = \$86,400$

14 The solid produced will be a prism with the cross-section shown in the figure. The first step is to calculate the area of the cross-section. Divide the cross-section into the 5 regions shown below. The 5th region represents the hole and must be subtracted from the other areas.

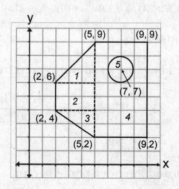

$A_1 = \frac{1}{2}b \cdot h$	$A_2 = l \cdot w$	$A_3 = \frac{1}{2}b \cdot h$	$A_4 = l \cdot w$	$A_5 = \pi R^2$
$= \frac{1}{2} \cdot 3 \cdot 3$	$= 3 \cdot 2$	$= \frac{1}{2} \cdot 3 \cdot 2$	$= 4 \cdot 7$	$= \pi(1)^2$
$= 4.5$	$= 6$	$= 3$	$= 28$	$= 3.14159$

$$A_{\text{drawing}} = A_1 + A_2 + A_3 + A_4 - A_5$$

$$= 4.5 + 6 + 3 + 28 - 3.14159 = 38.35841 \text{ units}^2$$

To find the cross-sectional area of the actual part in centimeters, use the scale factor and the fact that area is proportional to the scale factor *squared*.

$$A_{\text{cross-section}} = 38.35841 \text{ units}^2 \cdot \left(\frac{2 \text{ cm}}{1 \text{ unit}}\right)^2$$

$$= 153.43364 \text{ cm}^2$$

The solid is formed by translating the cross-section 10 units, or 20 cm. Use these dimensions to find the volume of the prism.

$$V = B \times h \qquad \text{where } B \text{ is the cross-sectional area}$$

$$= 153.43364 \text{ cm}^2 \cdot 20 \text{ cm}$$

$$= 3{,}068.6728 \text{ cm}^3$$

Now calculate the mass of the metal

$$= 8.44 \frac{\text{grams}}{\text{cm}^3} \cdot 3{,}068.6728 \text{ cm}^3 \qquad \text{mass} = \text{density} \times \text{volume}$$

$$= 25{,}899.5984 \text{ grams}$$

The price is given per kilogram, so convert grams to kilograms.

$$\text{Mass} = 25{,}899.5984 \text{ grams} \cdot \frac{1 \text{ kg}}{1{,}000 \text{ grams}}$$

$$= 25.899 \text{ kg}$$

Finally, find the cost using the price per kilogram.

$$\text{Cost} = 25.899 \text{ kg} \cdot \$32/\text{kg} \qquad \text{total cost} = \text{mass (price per kg)}$$

$$= \$829$$

Glossary of Geometry Terms

Acute angle An angle whose measure is greater than 0° and less than 90°.

Acute triangle A triangle whose angles are all acute angles.

Adjacent angles Two angles that share a common vertex and one side, but do not share interior points.

Altitude (of a triangle) A segment from a vertex of a triangle perpendicular to the opposite side.

Angle A figure formed by two rays with a common endpoint. Symbol is $\angle$.

Angle bisector A line, segment, or ray that divides an angle into two congruent angles.

Angle measure The amount of opening of an angle, measured in degrees or radians.

Angle of depression The angle formed by the horizontal and the line of sight when looking downward to an object.

Angle of elevation The angle formed by the horizontal and the line of sight when looking upward to an object.

Angle of rotation The angle measure by which a figure or point spins around a center point.

Arc A portion of a circle, with two endpoints on the circle. A **major arc** measures more than 180°, a **minor arc** measures less than 180°, and a **semicircle** measures 180°. Symbol $\frown$.

Arc length The distance between the endpoints of an arc. Arc length equals $R\theta$ where θ is the arc measure in radians and R is the radius.

Arc measure The angle measure of an arc equal to the measure of the central angle that intercepts the arc.

Base (of a circular cone) The circular face of a cone; it is opposite the apex.

Base (of an isosceles triangle) The noncongruent side of an isosceles triangle.

Base (of a pyramid) The polygonal face of a pyramid that is opposite the apex.

Bases (of a prism) A pair of faces of a prism that are parallel, congruent polygons.

Bases (of a circular cylinder) The pair of congruent, parallel circular faces of a cylinder.

Base angles The angles formed at each end of the base of an isosceles triangle.

Bisector, or segment bisector A line, segment, or ray that passes through the midpoint of a segment.

Biconditional A compound statement in the form "If and only if *hypothesis* then *conclusion*." The hypothesis and conclusion are statements that can be true or false. The biconditional is true when the hypothesis and conclusion have the same truth value.

Center of dilation The fixed reference point used to determine the expansion or contraction of lengths in a dilation. The center of dilation is the only invariant point in a dilation.

Center of a regular polygon The center of the inscribed or circumscribed circle of a regular polygon.

Center–radius equation of a circle A circle can be represented on the coordinate plane by $(x - h)^2 + (y - k)^2 = r^2$, where the point (h, k) is the center of the circle and r is the radius.

Central angle An angle in a circle formed by two distinct radii.

Centroid of a triangle The point of concurrency of the three medians of a triangle.

Chord A segment whose endpoints lie on a circle.

Circle The set of points that are a fixed distance from a fixed center point. Symbol ⊙.

Circumcenter The point that is the center of the circle circumscribed about a polygon, equidistant from the vertices of a polygon, and the point of concurrency of the perpendicular bisectors of a triangle.

Circumference The distance around a circle. Circumference = $2\pi \cdot$ radius.

Coincide (coincident) Figures that lay entirely one on the other.

Collinear Points that lie on the same line.

Compass A tool for drawing accurate circles and arcs.

Complementary angles Angles whose measures sum to 90°.

Completing the square A method used to rewrite a quadratic expression of the form $ax^2 + bx + c$ as a squared binomial of the form $(dx + e)^2$.

Composition of transformations A sequence of transformations in a specified order.

Concave polygon A polygon with at least one diagonal outside the polygon.

Concentric circles Circles with the same center.

Concurrent When three or more lines all intersect at a single point.

Cone (circular) A solid figure with a single circular base, and a curved lateral face that tapers to a point called the apex.

Congruent Figures with the same size and shape. Symbol $\cong$.

Construction A figure drawn with only a compass and straightedge.

Convex polygon A polygon whose diagonals all lie within the polygon.

Coplanar Figures that lie in the same plane.

Corresponding parts A pair of parts (usually points, sides, or angles) of two figures that are paired together through a specified relationship, such as a congruence or similarity statement or a transformation function.

Cosine of an angle In a right triangle, the ratio of the length of the side adjacent to an acute angle to the length of the hypotenuse.

CPCTC Corresponding parts of congruent triangles are congruent.

Cross-section The intersection of a plane with a solid.

Cube A prism whose faces are all squares.

Cubic unit The amount of space occupied by a cube 1 unit of length in each dimension. (1 ft^3 is the amount of space occupied by a 1 ft $\times$ 1 ft $\times$ 1 ft cube.)

Cylinder The solid figure formed when a rectangle is rotated 360° about one of its sides.

Cylinder (circular) A solid with two parallel, congruent circular bases and a curved lateral face.

Degree A unit of angle measure equal to $\frac{1}{360}$ th of a complete rotation.

Diagonal A segment in a polygon whose endpoints are nonconsecutive vertices of the polygon.

Diameter A chord that passes through the center of a circle.

Dilation A similarity transformation about a center point O, which maps pre-image P to image P' such that O, P, and P' are collinear and $OP' = k \cdot OP$. Notation is D_k.

Direct transformation A transformation that preserves orientation.

Distance from a point to a line The length of the segment from the point perpendicular to the line.

Edge of a solid The intersection of two polygonal faces of a solid.

Endpoint The two points that bound a segment.

Equiangular A figure whose angles all have the same measure.

Equidistant The same distance from two or more points.

Equilateral A figure whose sides all have the same length.

Exterior angle of a polygon The angle formed by a side of a polygon and the extension of an adjacent side.

Face of a solid Any of the surfaces that bound a solid.

Figure Any set of points. The points may form a segment, line, ray, plane, polygon, curve, solid, etc.

Glide reflection The composition of a line reflection and a translation along a vector parallel to the line of reflection.

Great circle The largest circle that can be drawn on a sphere. The circle formed by the intersection of a sphere and a plane passing through the center of the sphere.

Hexagon A 6-sided polygon.

Hypotenuse The longest side of a right triangle; it is always opposite the right angle.

Identity transformation A transformation in which the preimage and image are coincident.

Image The figure that results from applying a transformation to an initial figure called the preimage.

Incenter of a triangle The point that is the center of the inscribed circle of a triangle, equidistant from the 3 sides of a triangle, and the point of concurrency of the three angle bisectors of a triangle.

Inscribed angle An angle in a circle formed by two chords with a common endpoint.

Inscribed circle of a triangle A circle that is tangent to all three sides of the triangle.

Intersecting Figures that share at least one common point.

Isometry See *Rigid motion*.

Isosceles trapezoid A trapezoid with congruent legs.

Isosceles triangle A triangle with at least two congruent sides.

Kite A polygon with two distinct pairs of adjacent congruent sides. The opposite sides are not congruent.

Lateral edge The intersection between two lateral faces of a polyhedron.

Lateral face Any face of a polyhedron that is not a base.

Line One of the undefined terms in geometry. An infinitely long set of points that has no width or thickness. Symbol ↔.

Linear pair of angles Two adjacent supplementary angles.

Line symmetry A line over which one half of a figure can be reflected and mapped onto the other half of the figure.

Line reflection A rigid motion in which every point P is transformed to a point P' such that the line is the perpendicular bisector of $\overline{PP'}$.

Major arc An arc whose degree measure is greater than 180°.

Map (**mapping**) A pairing of every point in a preimage with a point in an image. Reflections, rotations, translations, and dilations are one-to-one mappings because every point in the preimage maps to exactly one point in the image, and every point in the image maps to exactly one point in the preimage.

Mean proportional (**geometric mean**) The square root of the product of two numbers, a and b. If $\dfrac{a}{m} = \dfrac{m}{b}$, then m is the geometric mean.

Median of a triangle The segment from a vertex of a triangle to the midpoint of the opposite side.

Midpoint A point that divides a segment into two congruent segments.

Midsegment (median) of a trapezoid A segment joining the midpoints of the two legs of a trapezoid.

Midsegment of a triangle A segment joining the midpoints of two sides of a triangle.

Minor arc An arc whose degree measure is less than 180°.

Noncollinear Points that do not lie on the same line.

Noncoplanar Points or lines that do not lie on the same plane.

Obtuse angle An angle whose measure is greater than 90° and less than 180°.

Octagon A polygon with 8 sides.

Opposite rays Two rays with a common endpoint that together form a straight line.

Opposite transformation A transformation that changes the orientation of a figure.

Orientation The order in which the points of a figure are encountered when moving around a figure. The orientation can be clockwise or counterclockwise.

Orthocenter The point of concurrence of the three altitudes of a triangle.

Parallel lines Coplanar lines that do not intersect. Symbol ‖.

Parallelogram A quadrilateral with two pairs of opposite parallel sides.

Parallel planes Planes that do not intersect.

Pentagon A polygon with 5 sides.

Perimeter The sum of the lengths of the sides of a polygon.

Perpendicular Intersecting at right angles. Symbol $\perp$.

Perpendicular bisector A line, segment, or ray perpendicular to another segment at its midpoint.

Pi The ratio of the circumference of a circle to its diameter. Pi is an irrational number whose value is approximately 3.14159. Symbol π.

Plane One of the undefined terms in geometry. A set of points with no thickness that extends infinitely in all directions. It is usually visualized as a flat surface.

Point An undefined term in geometry. A location in space with no length, width, or thickness. Symbol $\bullet$.

Point of tangency The point where a tangent intersects a curve.

Point reflection A transformation in which a specified center point is the midpoint of each of the segments connecting any point in the preimage with its corresponding point in the image. It is equivalent to a rotation of 180°.

Point-slope equation of a line A form of the equation of a line, $y - y_1 = m(x - x_1)$, where m is the slope and (x_1, y_1) are the coordinates of a point on the line.

Polygon A closed planar figure whose sides are segments that intersect only at their endpoints (do not overlap).

Polyhedron A solid figure in which each face is a polygon. Plural *polyhedra*. Prisms and pyramids are examples of polyhedra.

Postulate A statement that is accepted to be true without proof.

Preimage The original figure that is acted on by a transformation.

Preserves Remains unchanged.

Prism A polyhedron with two congruent, parallel polygons for bases, and whose lateral faces are parallelograms.

Proportion An equation that states two ratios are equal. For example, $\frac{a}{b} = \frac{c}{d}$. The cross-products of a proportion are equal, $a \cdot d = b \cdot c$.

Pyramid A polyhedron having one polygonal base and triangles for lateral faces.

Pythagorean theorem In a right triangle, the sum of the squares of the two legs equals the square of the hypotenuse, or $a^2 + b^2 = c^2$, where a and b are the lengths of the two legs and c is the length of the hypotenuse.

Quadratic equation An equation in the form $ax^2 + bx + c = 0$, where $a \neq 0$.

Quadratic formula A formula for finding the two solutions to a quadratic equation of the form $ax^2 + bx + c = 0$, $x = \dfrac{-b \pm \sqrt{b^2 - 4ac}}{2a}$.

Quadrilateral A polygon with four sides.

Radian An angle measure in which one full rotation is 2π radians. Also, 1 radian is the measure of an arc such that the arc's length is equal to the radius of that circle.

Radius A segment from the center of a circle to a point on the circle.

Ray A portion of a line starting at an endpoint and including all points on one side of the endpoint. Symbol $\rightarrow$.

Rectangle A parallelogram with right angles.

Reflection See *Line reflection* and *Point reflection*.

Reflexive property of equality Any quantity is equal to itself. Also, for figures, any figure is congruent to itself.

Regular polygon A polygon in which all sides are congruent and all angles are congruent.

Rhombus A parallelogram with 4 congruent sides.

Right angle An angle that measures 90°.

Right circular cone A cone with a circular base and whose altitude passes through the center of the base.

Right circular cylinder A cylinder with circular bases and whose altitude passes through the center of the bases.

Right pyramid A pyramid whose faces are isosceles triangles.

Right triangle A triangle that contains a right angle.

Rigid motion (isometry) A transformation that preserves distance. The image and preimage are congruent under a rigid motion. Translations, reflections, and rotations are isometries.

Rotation A rigid motion in which every point, P, in the preimage spins by a fixed angle around a center point, C, to point P'. The distance to the center point is preserved.

Rotational symmetry A figure has rotational symmetry if it maps onto itself after a rotation of less than 360°.

Scale drawing A drawing of a figure or object that represents a dilation of the actual figure or object. A drawing of an object in which every length is enlarged or reduced by the same scale factor.

Scale factor The ratio by which a figure is enlarged or reduced by a dilation.

Scalene triangle A triangle in which no sides have the same length.

Secant of a circle A line that intersects a circle in exactly two points.

Sector of a circle A region bounded by a central angle of a circle and the arc it intersects.

Segment A portion of a line bounded by two endpoints.

Semicircle An arc that measures 180°.

Similarity transformation A transformation in which the preimage and image are similar. A transformation that includes a dilation.

Similar polygons Two polygons with the same shape but not necessarily the same size.

Sine of an angle In a right triangle, the ratio of the length of the side opposite an acute angle to the length of the hypotenuse.

Skew lines Two lines that are not coplanar.

Slant height The distance along a lateral face of a solid from the apex perpendicular to the opposite edge.

Slope A numerical measure of the steepness of a line. In the coordinate plane, the slope of a line equals the change in the y-coordinates divided by the change in the x-coordinates between any two points. The slope of a vertical line is undefined.

Slope-intercept equation of a line A form of the equation of a line, $y = mx + b$, where m is the slope and b is the y-intercept.

Solid figure A 3-dimensional figure fully enclosed by surfaces.

Sphere A solid comprised of the set of points in space that are a fixed distance from a center point.

Square A parallelogram with 4 right angles and 4 congruent sides.

Straightedge A ruler with no length marking, used for constructing straight lines.

Substitution property of equality The property that a quantity can be replaced by an equal quantity in an equation.

Subtraction property of equality If the same or equal quantities are subtracted from the same or equal quantities, then the differences are equal.

Supplementary angles Two angles whose measures sum to 180°.

Surface area The sum of the areas of all the faces or curved surfaces of a solid figure.

Tangent of an angle In a right triangle, the ratio of the length of the side opposite an acute angle to the length of the side adjacent to the angle.

Tangent to a circle A line, coplanar with a circle, that intersects the circle at only one point.

Theorem A general statement that can be proven.

Transformation A one-to-one function that maps a set of points, called the preimage, to a new set of points, called the image.

Transitive property of congruence The property that states: if figure $A \cong$ figure B and figure $B \cong$ figure C, then figure $A \cong$ figure C.

Transitive property of equality The property that states: if $a = b$ and $b = c$, then $a = c$.

Translation A rigid motion that slides every point in the preimage in the same direction and by the same distance.

Transversal A line that intersects two or more other lines at different points.

Trapezoid A quadrilateral that has at least one pair of parallel sides.

Undefined terms Terms that cannot be formally defined using previously defined terms. In geometry these traditionally include point, line, and plane.

Vector A quantity that has both magnitude and direction; represented geometrically by a directed line segment. Symbol $\rightarrow$.

Vertex The point of intersection of two consecutive sides of a polygon or the two rays of an angle.

Vertex angle The angles formed by the two congruent sides in an isosceles triangle.

Vertical angles The pairs of opposite angles formed by two intersecting lines.

Volume The amount of space occupied by a solid figure measured in cubic units (in^3, cm^3, etc.). The number of nonoverlapping unit cubes that can fit in the interior of a solid.

Zero product property The property that states: if $a \cdot b = 0$, then either $a = 0$, or $b = 0$, or a and $b = 0$.

Regents Exams, Answers, and Self-Analysis Charts

June 2019 Exam
Geometry

HIGH SCHOOL MATH REFERENCE SHEET

Conversions

1 inch = 2.54 centimeters

1 meter = 39.37 inches

1 mile = 5280 feet

1 mile = 1760 yards

1 mile = 1.609 kilometers

1 kilometer = 0.62 mile

1 pound = 16 ounces

1 pound = 0.454 kilogram

1 kilogram = 2.2 pounds

1 ton = 2000 pounds

1 cup = 8 fluid ounces

1 pint = 2 cups

1 quart = 2 pints

1 gallon = 4 quarts

1 gallon = 3.785 liters

1 liter = 0.264 gallon

1 liter = 1000 cubic centimeters

Formulas

Triangle	$A = \frac{1}{2}bh$
Parallelogram	$A = bh$
Circle	$A = \pi r^2$
Circle	$C = \pi d$ or $C = 2\pi r$

Formulas (continued)

General Prism	$V = Bh$
Cylinder	$V = \pi r^2 h$
Sphere	$V = \frac{4}{3}\pi r^3$
Cone	$V = \frac{1}{3}\pi r^2 h$
Pyramid	$V = \frac{1}{3}Bh$
Pythagorean Theorem	$a^2 + b^2 = c^2$
Quadratic Formula	$x = \dfrac{-b \pm \sqrt{b^2 - 4ac}}{2a}$
Arithmetic Sequence	$a_n = a_1 + (n - 1)d$
Geometric Sequence	$a_n = a_1 r^{n-1}$
Geometric Series	$S_n = \dfrac{a_1 - a_1 r^n}{1 - r}$ where $r \neq 1$
Radians	$1 \text{ radian} = \dfrac{180}{\pi} \text{ degrees}$
Degrees	$1 \text{ degree} = \dfrac{\pi}{180} \text{ radians}$
Exponential Growth/Decay	$A = A_0 e^{k(t - t_0)} + B_0$

180 Minutes–35 Questions

PART I

Answer all 24 questions in this part. Each correct answer will receive 2 credits. No partial credit will be allowed. For each statement or question, write in the space provided the numeral preceding the word or expression that best completes the statement or answers the question. [48 credits]

1 On the set of axes below, triangle ABC is graphed. Triangles $A'B'C'$ and $A''B''C''$, the images of triangle ABC, are graphed after a sequence of rigid motions.

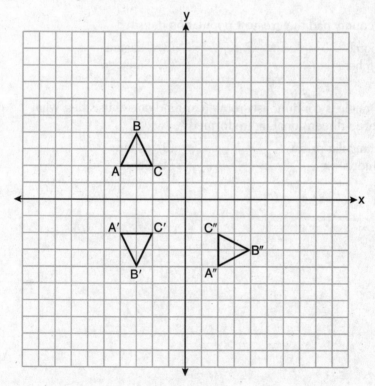

Identify which sequence of rigid motions maps $\triangle ABC$ onto $\triangle A'B'C'$ and then maps $\triangle A'B'C'$ onto $\triangle A''B''C''$.

(1) a rotation followed by another rotation
(2) a translation followed by a reflection
(3) a reflection followed by a translation
(4) a reflection followed by a rotation

1 _____

2 The table below shows the population and land area, in square miles, of four counties in New York State at the turn of the century.

County	2000 Census Population	2000 Land Area (mi²)
Broome	200,536	706.82
Dutchess	280,150	801.59
Niagara	219,846	522.95
Saratoga	200,635	811.84

Which county had the greatest population density?

(1) Broome   (3) Niagara
(2) Dutchess   (4) Saratoga   2 _____

3 If a rectangle is continuously rotated around one of its sides, what is the three-dimensional figure formed?

(1) rectangular prism   (3) sphere
(2) cylinder   (4) cone   3 _____

4 Which transformation carries the parallelogram below onto itself?

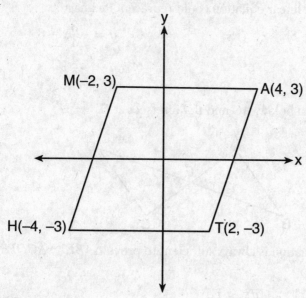

(1) a reflection over $y = x$
(2) a reflection over $y = -x$
(3) a rotation of 90° counterclockwise about the origin
(4) a rotation of 180° counterclockwise about the origin   4 _____

5 After a dilation centered at the origin, the image of $\overline{CD}$ is $\overline{C'D'}$. If the coordinates of the endpoints of these segments are $C(6,-4)$, $D(2,-8)$, $C'(9,-6)$, and $D'(3,-12)$, the scale factor of the dilation is

(1) $\dfrac{3}{2}$ (3) 3

(2) $\dfrac{2}{3}$ (4) $\dfrac{1}{3}$ 5 _____

6 A tent is in the shape of a right pyramid with a square floor. The square floor has side lengths of 8 feet. If the height of the tent at its center is 6 feet, what is the volume of the tent, in cubic feet?

(1) 48   (3) 192
(2) 128   (4) 384   6 _____

7 The line $-3x + 4y = 8$ is transformed by a dilation centered at the origin. Which linear equation could represent its image?

(1) $y = \frac{4}{3}x + 8$

(3) $y = -\frac{3}{4}x - 8$

(2) $y = \frac{3}{4}x + 8$

(4) $y = -\frac{4}{3}x - 8$

7 _____

8 In the diagram below, $\overline{AC}$ and $\overline{BD}$ intersect at E.

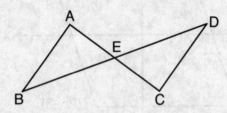

Which information is always sufficient to prove $\triangle ABE \cong \triangle CDE$?

(1) $\overline{AB} \parallel \overline{CD}$

(2) $\overline{AB} \cong \overline{CD}$ and $\overline{BE} \cong \overline{DE}$

(3) E is the midpoint of $\overline{AC}$.

(4) $\overline{BD}$ and $\overline{AC}$ bisect each other.

8 _____

9 The expression $\sin 57°$ is equal to

(1) $\tan 33°$

(3) $\tan 57°$

(2) $\cos 33°$

(4) $\cos 57°$

9 _____

10 What is the volume of a hemisphere that has a diameter of 12.6 cm, to the *nearest tenth of a cubic centimeter*?

(1) 523.7

(3) 4189.6

(2) 1047.4

(4) 8379.2

10 _____

11 In the diagram below of $\triangle ABC$, D is a point on $\overline{BA}$, E is a point on $\overline{BC}$, and $\overline{DE}$ is drawn.

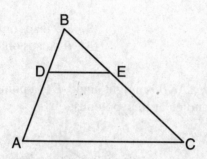

If $BD = 5$, $DA = 12$, and $BE = 7$, what is the length of $\overline{BC}$ so that $\overline{AC} \parallel \overline{DE}$?

(1) 23.8   (3) 15.6
(2) 16.8   (4) 8.6   11 _____

12 A quadrilateral must be a parallelogram if
  (1) one pair of sides is parallel and one pair of angles is congruent
  (2) one pair of sides is congruent and one pair of angles is congruent
  (3) one pair of sides is both parallel and congruent
  (4) the diagonals are congruent   12 _____

13 In the diagram below of circle O, chords $\overline{JT}$ and $\overline{ER}$ intersect at M.

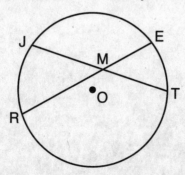

If $EM = 8$ and $RM = 15$, the lengths of $\overline{JM}$ and $\overline{TM}$ could be

(1) 12 and 9.5   (3) 16 and 7.5
(2) 14 and 8.5   (4) 18 and 6.5   13 _____

14 Triangles *JOE* and *SAM* are drawn such that $\angle E \cong \angle M$ and $\overline{EJ} \cong \overline{MS}$. Which mapping would *not* always lead to $\triangle JOE \cong \triangle SAM$?

(1) $\angle J$ maps onto $\angle S$ (3) $\overline{EO}$ maps onto $\overline{MA}$

(2) $\angle O$ maps onto $\angle A$ (4) $\overline{JO}$ maps onto $\overline{SA}$ 14 _____

15 In $\triangle ABC$ shown below, $\angle ACB$ is a right angle, E is a point on $\overline{AC}$, and $\overline{ED}$ is drawn perpendicular to hypotenuse $\overline{AB}$.

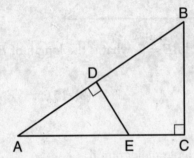

If $AB = 9$, $BC = 6$, and $DE = 4$, what is the length of $\overline{AE}$?

(1) 5   (3) 7

(2) 6   (4) 8   15 _____

16 Which equation represents a line parallel to the line whose equation is $-2x + 3y = -4$ and passes through the point (1,3)?

(1) $y - 3 = -\frac{3}{2}(x - 1)$ (3) $y + 3 = -\frac{3}{2}(x + 1)$

(2) $y - 3 = \frac{2}{3}(x - 1)$ (4) $y + 3 = \frac{2}{3}(x + 1)$ 16 _____

17 In rhombus *TIGE*, diagonals $\overline{TG}$ and $\overline{IE}$ intersect at *R*. The perimeter of *TIGE* is 68, and *TG* = 16.

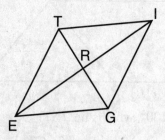

What is the length of diagonal $\overline{IE}$?

(1) 15

(2) 30

(3) 34

(4) 52

17 _____

18 In circle *O* two secants, $\overline{ABP}$ and $\overline{CDP}$, are drawn to external point *P*. If $m\widehat{AC} = 72°$, and $m\widehat{BD} = 34°$, what is the measure of $\angle P$?

(1) 19°

(2) 38°

(3) 53°

(4) 106°

18 _____

19 What are the coordinates of point *C* on the directed segment from *A*(−8,4) to *B*(10,−2) that partitions the segment such that *AC* : *CB* is 2 : 1?

(1) (1,1)

(2) (−2,2)

(3) (2,−2)

(4) (4,0)

19 _____

20 The equation of a circle is $x^2 + 8x + y^2 - 12y = 144$. What are the coordinates of the center and the length of the radius of the circle?

(1) center (4,−6) and radius 12

(2) center (−4,6) and radius 12

(3) center (4,−6) and radius 14

(4) center (−4,6) and radius 14

20 _____

21 In parallelogram $PQRS$, $\overline{QP}$ is extended to point T and $\overline{ST}$ is drawn.

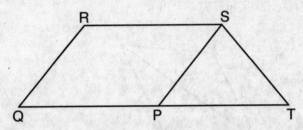

If $\overline{ST} \cong \overline{SP}$ and m∠$R = 130°$, what is m∠PST?

(1) 130°   (3) 65°
(2) 80°   (4) 50°   21 _____

22 A 12-foot ladder leans against a building and reaches a window
    10 feet above ground. What is the measure of the angle, to the
    nearest degree, that the ladder forms with the ground?

(1) 34   (3) 50
(2) 40   (4) 56   22 _____

23 In the diagram of equilateral triangle ABC shown below, E and F
    are the midpoints of $\overline{AC}$ and $\overline{BC}$, respectively.

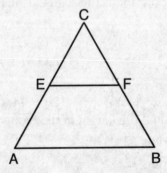

If $EF = 2x + 8$ and $AB = 7x - 2$, what is the perimeter of
trapezoid $ABFE$?

(1) 36   (3) 100
(2) 60   (4) 120   23 _____

24 Which information is *not* sufficient to prove that a parallelogram is a square?

  (1) The diagonals are both congruent and perpendicular.

  (2) The diagonals are congruent and one pair of adjacent sides are congruent.

  (3) The diagonals are perpendicular and one pair of adjacent sides are congruent.

  (4) The diagonals are perpendicular and one pair of adjacent sides are perpendicular.   24 _____

PART II

**Answer all 7 questions in this part. Each correct answer will receive
2 credits. Clearly indicate the necessary steps, including appropriate
formula substitutions, diagrams, graphs, charts, etc. For all questions in
this part, a correct numerical answer with no work shown will receive
only 1 credit.** [14 credits]

25 Triangle $A'B'C'$ is the image of triangle ABC after a dilation with a
scale factor of $\frac{1}{2}$ and centered at point A. Is triangle ABC
congruent to triangle $A'B'C'$? Explain your answer.

26 Determine and state the area of triangle PQR, whose vertices have coordinates $P(-2,-5)$, $Q(3,5)$, and $R(6,1)$.

[The use of the set of axes below is optional.]

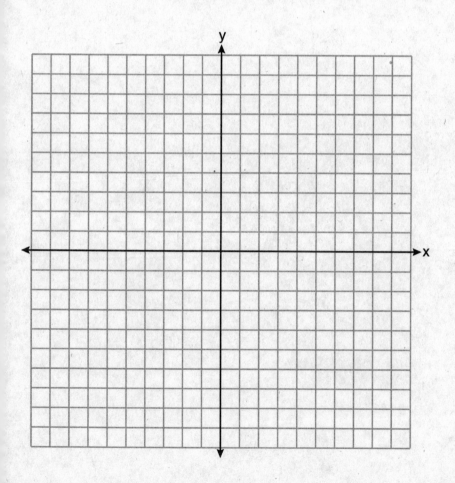

27 A support wire reaches from the top of a pole to a clamp on the ground. The pole is perpendicular to the level ground and the clamp is 10 feet from the base of the pole. The support wire makes a 68° angle with the ground. Find the length of the support wire to the *nearest foot*.

28 In the diagram below, circle O has a radius of 10.

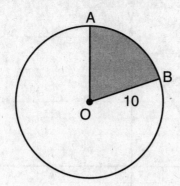

If $m\widehat{AB} = 72°$, find the area of shaded sector AOB, in terms of π.

29 On the set of axes below, $\triangle ABC \cong \triangle STU$.

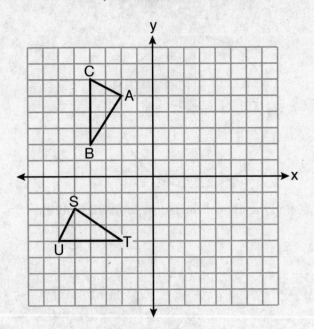

Describe a sequence of rigid motions that maps $\triangle ABC$ onto $\triangle STU$.

30 In right triangle *PRT*, m∠*P* = 90°, altitude $\overline{PQ}$ is drawn to hypotenuse $\overline{RT}$, *RT* = 17, and *PR* = 15.

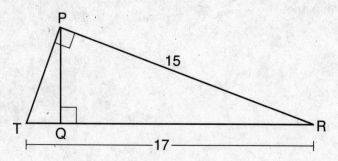

Determine and state, to the *nearest tenth*, the length of $\overline{RQ}$.

31 Given circle O with radius $\overline{OA}$, use a compass and straightedge to construct an equilateral triangle inscribed in circle O. [Leave all construction marks.]

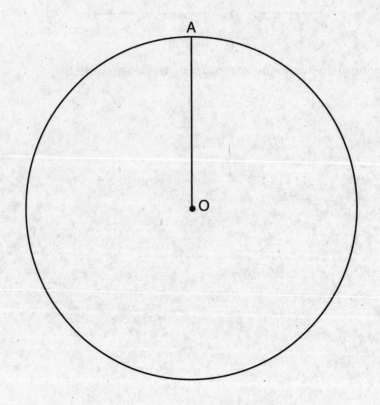

PART III

Answer all 3 questions in this part. Each correct answer will receive 4 credits. Clearly indicate the necessary steps, including appropriate formula substitutions, diagrams, graphs, charts, etc. Utilize the information provided for each question to determine your answer. Note that diagrams are not necessarily drawn to scale. For all questions in this part, a correct numerical answer with no work shown will receive only 1 credit.   [12 credits]

32 Riley plotted $A(-1,6)$, $B(3,8)$, $C(6,-1)$, and $D(1,0)$ to form a quadrilateral. Prove that Riley's quadrilateral $ABCD$ is a trapezoid.
[The use of the set of axes on the next page is optional.]

Question 32 is continued on the next page.

Question 32 continued

Riley defines an isosceles trapezoid as a trapezoid with congruent diagonals. Use Riley's definition to prove that *ABCD* is *not* an isosceles trapezoid.

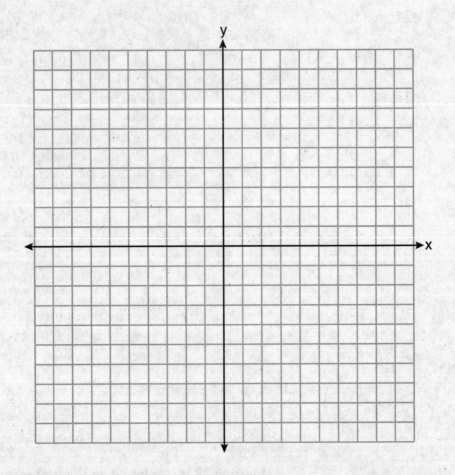

33 A child-sized swimming pool can be modeled by a cylinder. The pool has a diameter of $6\frac{1}{2}$ feet and a height of 12 inches. The pool is filled with water to $\frac{2}{3}$ of its height. Determine and state the volume of the water in the pool, to the *nearest cubic foot*.

One cubic foot equals 7.48 gallons of water. Determine and state, to the *nearest gallon*, the number of gallons of water in the pool.

34 Nick wanted to determine the length of one blade of the windmill pictured below. He stood at a point on the ground 440 feet from the windmill's base. Using surveyor's tools, Nick measured the angle between the ground and the highest point reached by the top blade and found it was 38.8°. He also measured the angle between the ground and the lowest point of the top blade, and found it was 30°.

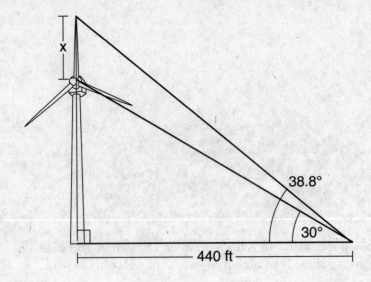

Determine and state a blade's length, x, to the *nearest foot*.

PART IV

Answer the question in this part. Each correct answer will receive 6 credits. Clearly indicate the necessary steps, including appropriate formula substitutions, diagrams, graphs, charts, etc. For all questions in this part, a correct numerical answer with no work shown will receive only 1 credit. [6 credits]

35 Given: Quadrilateral $MATH$, $\overline{HM} \cong \overline{AT}$, $\overline{HT} \cong \overline{AM}$, $\overline{HE} \perp \overline{MEA}$, and $\overline{HA} \perp \overline{AT}$

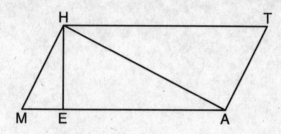

Prove: $TA \cdot HA = HE \cdot TH$

Work space for question 35 is continued on the next page.

Question 35 continued

Answers June 2019
Geometry

Answer Key

PART I

1. 4	5. 1	9. 2	13. 3	17. 2	21. 2
2. 3	6. 2	10. 1	14. 4	18. 1	22. 4
3. 2	7. 2	11. 1	15. 2	19. 4	23. 3
4. 4	8. 4	12. 3	16. 2	20. 4	24. 3

PART II

25. The triangles are not congruent because corresponding sides are not congruent.

26. area = 25

27. length = 27 ft

28. area = 20π

29. Rotate 90° counterclockwise about the origin.

30. $RQ = 13.2$

31. See the detailed solution for the construction.

PART III

32. $ABCD$ is a trapezoid because $\overline{AD} \parallel \overline{BC}$.
$ABCD$ is not isosceles because $\overline{AC}$ is not congruent to $\overline{BD}$.

33. volume = 22 ft³, number of gallons = 165

34. length = 100 ft

PART IV

35. See the detailed solution for the proof.

In **PARTS II–IV,** you are required to show how you arrived at your answers. For sample methods of solutions, see the *Answers Explained* section.

Answers Explained

PART I

1. The sequence of transformations needs to map the following points:

$$A(-4, 2) \rightarrow A'(-4, -2) \rightarrow A''(2, -4)$$

   A reflection over the x-axis changes the sign of the y-coordinate, and takes A to A'. A counterclockwise rotation of $90°$ centered at the origin changes coordinates (x, y) to $(-y, x)$, and takes A' to A''. This same sequence also maps B to B'' and C to C''.

   The correct choice is (**4**).

2. Find the population density of each county by dividing the population by the area.

   Broome: population density $= \dfrac{200{,}536}{706.82} = 283.716$

   Dutchess: population density $= \dfrac{280{,}150}{801.59} = 349.493$

   Niagara: population density $= \dfrac{219{,}846}{522.95} = 428.396$

   Saratoga: population density $= \dfrac{200{,}635}{811.84} = 247.136$

   Niagara County has the greatest population density.

   The correct choice is (**3**).

3. The solid formed by rotating a rectangle about one of its sides is a cylinder.

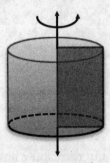

   The correct choice is (**2**).

4. One approach is to choose a vertex and apply each transformation. We are looking to map the chosen vertex to one of the other vertices. If the vertex does not map to one of the original vertices, that transformation can be eliminated. Using vertex $M(-2, 3)$ we apply each transformation.

Choice 1 – A reflection over the line $y = x$ swaps the coordinates with the rule $(x, y) \rightarrow (y, x)$. $M(-2, 3) \rightarrow M'(3, -2)$, which is not a vertex, so this choice is eliminated.

Choice 2 – A reflection over the line $y = -x$ swaps the coordinates and changes the signs with the rule $(x, y) \rightarrow (-y, -x)$. $M(-2, 3) \rightarrow M'(-3, 2)$, which is not a vertex, so this choice is eliminated.

Choice 3 – A counterclockwise rotation of $90°$ about the origin transforms the coordinates with the rule $(x, y) \rightarrow (-y, x)$. $M(-2, 3) \rightarrow M'(-3, -2)$, which is not a vertex, so this choice is eliminated.

Choice 4 – A counterclockwise rotation of $180°$ about the origin transforms the coordinates with the rule $(x, y) \rightarrow (-x, -y)$. $M(-2, 3) \rightarrow M'(2, -3)$, which is vertex T. You should confirm that the same rule maps A to H, H to A, and T to M.

The correct choice is **(4)**.

5. A dilation always multiples each coordinate by the scale factor. Work backwards by dividing the new coordinate by the original to calculate the scale factor. We can use any point and either coordinate. Using the x-coordinates of points C and C' we find:

$$\text{scale factor} = \frac{x\text{-coordinate of } C'}{x\text{-coordinate of } C}$$

$$= \frac{9}{6}$$

$$= \frac{3}{2}$$

You can confirm this result using the other coordinates.

The correct choice is **(1)**.

6. The volume of a pyramid is given by $V = \frac{1}{3}Bh$ where B is the area of the base and h is the height. First find the area of the square base.

$$B = \text{side}^2$$
$$= 8^2$$
$$= 64$$

Now apply the volume formula with a height of 6.

$$V = \frac{1}{3}Bh$$
$$= \frac{1}{3}(64)(6)$$
$$= 128$$

The volume of the tent is 128 ft^2.

The correct choice is (**2**).

7. Dilating a line through the origin multiplies the y-intercept by the scale factor, but leaves the slope unchanged. The first step is to identify the slope of the original line by rewriting it in $y = mx + b$ form.

$$-3x + 4y = 8$$
$$4y = 3x + 8$$
$$y = \frac{3}{4}x + 2$$

The slope of the line is $\frac{3}{4}$ and the y-intercept is 2. A dilated line must also have a slope of $\frac{3}{4}$. The only choice with the correct slope is $y = \frac{3}{4}x + 8$.

The correct choice is (**2**).

8. The only congruent parts we are given are the vertical angles, $\angle AEB \cong \angle DEC$. Examine each of the choices and determine if it provides enough information to apply one of the triangle congruence postulates.

Choice 1 – Parallel sides $\overline{AB}$ and $\overline{CD}$ result in congruent alternate interior angles, so we can conclude $\angle A \cong \angle C$ and $\angle B \cong \angle D$. All three angles are congruent, but AAA is not a congruence postulate.

Choice 2 – $\overline{AB} \cong \overline{CD}$ and $\overline{BE} \cong \overline{DE}$ give 2 pairs of congruent sides, but the congruent angles are not the included angles. This choice results in ASS, which is not a congruence postulate.

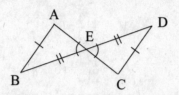

Choice 3 – Midpoint E lets us conclude $\overline{AE} \cong \overline{AC}$. This one additional pair of sides results in SA, which is not a congruence postulate.

Choice 4 – $\overline{BD}$ and $\overline{AC}$ bisecting each other implies E is the midpoint of $\overline{AC}$ and $\overline{BD}$.

Therefore, $\overline{BE} \cong \overline{ED}$ and $\overline{AE} \cong \overline{EC}$ for SAS, which is one of the triangle congruence postulates.

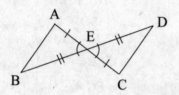

The correct choice is (**4**).

9. The co-function relationship states that if two angles A and B sum to $90°$, then $\sin(A) = \cos(B)$. Since $57° + 33° = 90°$, $\sin(57°)$ must equal $\cos(33°)$. Alternatively, you can use your calculator to evaluate each of the expressions. Both $\sin(57°)$ and $\cos(33°)$ are equal to 0.838671.

The correct choice is (**2**).

10. The radius of the hemisphere is found by dividing the 12.6 cm diameter by 2, so the radius is 6.3 cm. A hemisphere is half of a sphere, so multiply the volume formula of a sphere by $\frac{1}{2}$.

$$V = \frac{1}{2}\left(\frac{4}{3}\pi R^3\right)$$

$$= \frac{1}{2}\left(\frac{4}{3}\pi\left(6.3^3\right)\right)$$

$$= 523.7$$

The correct choice is **(1)**.

11. A segment parallel to a side of a triangle forms a triangle similar to the original. $\triangle BDE \sim \triangle BAC$, and we can use the side splitter theorem to solve for EC.

$$\frac{BD}{DA} = \frac{BE}{EC}$$

$$\frac{5}{12} = \frac{7}{EC}$$

$$5EC = 7 \cdot 12$$

$$EC = \frac{84}{5}$$

$$= 16.8$$

Now calculate BC.

$$BC = BE + EC$$

$$= 7 + 16.8$$

$$= 23.8$$

The correct choice is **(1)**.

12. The only choice that always specifies a parallelogram is choice (3). A counter-example can be identified for each of the other choices as shown in the figures.

     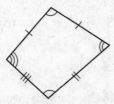   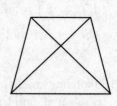

1 pair of parallel sides   1 pair of congruent sides   congruent diagonals

1 pair of congruent angles   1 pair of congruent angles

The correct choice is **(3)**.

13. When two chords intersect in a circle, the products of the parts of the chords are equal.

$$JM \cdot TM = EM \cdot RM$$
$$JM \cdot TM = 8 \cdot 15$$
$$JM \cdot TM = 120$$

Now check which pair has a product of 120. The only choice that works is $16 \cdot 7.5 = 120$.

The correct choice is **(3)**.

14. Sketch the two triangles with the given congruent parts marked. Check each choice to determine which additional pair of parts would not result in a valid triangle congruence postulate.

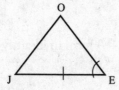   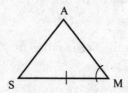

$\angle J \cong \angle S$ results in triangles congruent by ASA.

$\angle O \cong \angle A$ results in triangles congruent by AAS.

$\overline{EO} \cong \overline{MA}$ results in triangles congruent by SAS.

$\overline{JO} \cong \overline{SA}$ results in ASS, which is not a triangle congruence postulate.

The correct choice is **(4)**.

15. Sketching $\triangle ABC$ and $\triangle AED$ separately, we see that right angles C and D are congruent, and the shared angle A is also congruent. Therefore, the triangles are similar by the AA postulate. The similarity statement is $\triangle ABC \sim \triangle AED$. Corresponding sides of similar triangles are proportional, so we can set up a proportion and solve for AE.

$$\frac{AE}{AB} = \frac{ED}{BC}$$

$$\frac{AE}{9} = \frac{4}{6}$$

$$6AE = 9 \cdot 4$$

$$AE = 6$$

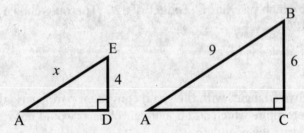

The correct choice is **(2)**.

16. First rewrite the original line in $y = mx + b$ form to identify the slope.

$$-2x + 3y = -4$$

$$3y = 2x - 4$$

$$y = \frac{2}{3}x - \frac{4}{3}$$

The slope of the original line is $\frac{2}{3}$. Parallel lines have equal slopes, so the new line must also have a slope of $\frac{2}{3}$. Write the equation of the parallel line using the point-slope form $y - y_1 = m(x - x_1)$.

Substitute $\frac{2}{3}$ for m and the given coordinates (1, 3) for x_1 and y_1 to get

$$y - 3 = \frac{2}{3}(x - 1).$$

The correct choice is **(2)**.

17. All sides of a rhombus are congruent, so divide the perimeter by 4 to find each side length.

$$\begin{aligned}
\text{side length} &= \frac{1}{4} \text{ perimeter} \\
&= \frac{68}{4} \\
&= 17
\end{aligned}$$

The diagonals of a rhombus bisect each other, so divide TG by 2 to find TR and RG.

$$TR = RG = \frac{16}{2}$$

$$TR = RG = 8$$

The diagonals of a rhombus are perpendicular, making $\triangle GRE$ a right triangle. Apply the Pythagorean theorem in $\triangle GRE$ to find RE.

$$\begin{aligned}
a^2 + b^2 &= c^2 \\
RE^2 + RG^2 &= EG^2 \\
RE^2 + 8^2 &= 17^2 \\
RE^2 + 64 &= 289 \\
RE^2 &= 225 \\
RE &= \sqrt{225} \\
&= 15
\end{aligned}$$

Diagonal $\overline{IE}$ is twice the length of $\overline{RE}$, so $IE = 30$.

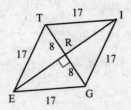

The correct choice is **(2)**.

18. The angle formed by two secants from an external point is equal to half the difference of the intercepted arcs.

$$m\angle P = \frac{1}{2}\left(m\widehat{AC} - m\widehat{BD}\right)$$

$$= \frac{1}{2}(72 - 34)$$

$$= 19$$

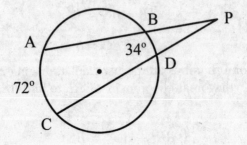

The correct choice is **(1)**.

19. There are several different methods for solving a directed line segment problem. A graphical approach is to sketch the starting and ending x-coordinates on a number line. Then using the given ratio, count jumps from the left and right. The two ends will meet at the desired x-coordinate. In this case the ratio is 2 : 1, so count two jumps from the left and 1 jump from the right.

After 6 jumps the two ends meet at an x-coordinate of 4. The only choice with an x-coordinate of 4 is (4, 0). If needed, you can repeat this procedure for the y-coordinate.

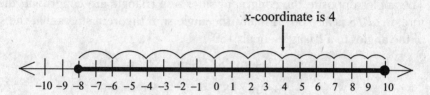

x-coordinate is 4

The correct choice is **(4)**.

20. Use the completing the square procedure to rewrite the equation in the form $(x - h)^2 + (y - k)^2 = r^2$.

First find the constant terms needed to complete the square:

The constant for the x-terms = (½ · coefficient of x)^2

$$= \left(\frac{1}{2} \cdot 8\right)^2$$

$$= (4)^2$$

$$= 16$$

The constant for the y-terms = (½ · coefficient of y)^2

$$= \left(\frac{1}{2} \cdot (-12)\right)^2$$

$$= (-6)^2$$

$$= 36$$

Add the required constant to each side of the equation and factor.

$$x^2 + 8x + 16 + y^2 - 12y + 36 = 144 + 16 + 36$$

$$(x + 4)^2 + (y - 6)^2 = 196$$

The values of h and k are -4 and 6. Be careful with the signs because h and k are the values subtracted from x and y. The center is located at $(-4, 6)$. Since $r^2 = 196$, the radius is equal to $\sqrt{196}$, or 14.

The correct choice is **(4)**.

21. Starting with m$\angle R = 130°$, we know that m$\angle QPS = 130°$ because opposite angles of a parallelogram are congruent. $\angle QPS$ and $\angle SPT$ are a linear pair that sum to 180°, making m$\angle SPT = 50°$. $\triangle SPT$ is isosceles with $ST = SP$. The angles opposite the congruent sides of a triangle are congruent; therefore, m$\angle PTS$ is also 50°. Finally, the angle sum theorem states that the sum of the angles in a triangle equals 180°.

$$\text{m}\angle PST + \text{m}\angle SPT + \text{m}\angle STP = 180°$$
$$\text{m}\angle PST + 50° + 50° = 180°$$
$$\text{m}\angle PST = 80°$$

The correct choice is **(2)**.

22. The ladder, building, and ground form a right triangle as shown. Trigonometry can be used to find the unknown angle using the two known sides. Relative to the unknown angle, the 12-foot ladder is the hypotenuse and the 10-foot height is the opposite. The opposite and hypotenuse indicate that a sine ratio is needed. After setting up the ratio, apply the inverse sine function to find the angle.

$$\sin(x) = \frac{\text{opposite}}{\text{hypotenuse}}$$
$$\sin(x) = \frac{10}{12}$$
$$x = \sin^{-1}\left(\frac{10}{12}\right)$$
$$= 56°$$

The correct choice is **(4)**.

23. Midpoint E and F form midsegment $\overline{EF}$, which is half the length of the opposite side $\overline{AB}$.

$$EF = \frac{1}{2}AB$$
$$2EF = AB$$
$$2(2x + 8) = 7x - 2$$
$$4x + 16 = 7x - 2$$
$$16 = 3x - 2$$
$$18 = 3x$$
$$x = 6$$

Substitute to find EF and AB.

$$EF = 2(6) + 8$$
$$= 20$$
$$AB = 7(6) - 2$$
$$= 40$$

Because $\triangle ABC$ is equilateral $AB = AC = 40$. AE is half of AC, so AE is 20. FB is also 20. Now add the side lengths to find the perimeter of $ABFE$.

$$\text{perimeter} = AE + EF + FB + AB$$
$$= 20 + 20 + 20 + 40$$
$$= 100$$

The correct choice is **(3)**.

24. To show a parallelogram is a square, you must show it has a special rectangle property and a special rhombus property. These properties are:

> rhombus – adjacent sides congruent, diagonals bisect each other, and diagonals bisect the vertex angles
> rectangle – congruent diagonals, and adjacent sides perpendicular

Perpendicular diagonals is a rectangle property, and congruent adjacent sides is a rhombus property.

The correct choice is **(3)**.

PART II

25. A dilation changes side lengths by multiplying them by the scale factor. Since all the side lengths of $\triangle A'BC'$ are half the side lengths of $\triangle ABC$, the triangles cannot be congruent. Congruent figures must have congruent corresponding sides.

26. The area of a triangle is found using $A = \frac{1}{2}bh$, where b is the base of the triangle and h is the height. $\triangle PQR$ appears to be a right triangle. We can confirm this by looking at the slopes of QR and PR. If they are negative reciprocals the sides are perpendicular, and the triangle is a right triangle.

$$\text{slope} = \frac{rise}{run}$$

$$\text{slope}_{QR} = -\frac{4}{3}$$

$$\text{slope}_{PR} = \frac{6}{8} = \frac{3}{4}$$

The two slopes are negative reciprocals, making $\triangle PQR$ a right triangle. $\overline{PR}$ is the base of the triangle and $\overline{QR}$ is the height. We can find the lengths using the distance formula.

$$d = \sqrt{\left(x_2 - x_1\right)^2 + \left(y_2 - y_1\right)^2}$$

$$QR = \sqrt{\left(6 - 3\right)^2 + \left(1 - 5\right)^2}$$

$$= \sqrt{\left(3\right)^2 + \left(-4\right)^2}$$

$$= \sqrt{9 + 16}$$

$$= 5$$

$$PR = \sqrt{\left(6 - (-2)\right)^2 + \left(1 - (-5)\right)^2}$$

$$= \sqrt{\left(8\right)^2 + \left(6\right)^2}$$

$$= \sqrt{64 + 36}$$

$$= 10$$

Now apply the area formula.

$$A = \frac{1}{2}bh$$

$$= \frac{1}{2}10 \cdot 5$$

$$= 25$$

The area of $\triangle PQR$ is 25.

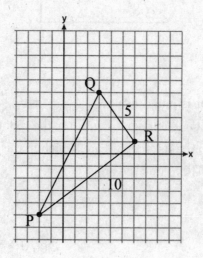

27. The wire, pole, and ground form a right triangle as shown. Relative to the 68° angle, the length of 10 along the ground is the adjacent and the unknown wire length is the hypotenuse. Therefore, we should use a cosine ratio to find the wire length.

$$\cos(68°) = \frac{10}{x}$$

$$x \cos(68°) = 10$$

$$x = \frac{10}{\cos(68)}$$

$$= 26.694$$

Rounded to the nearest foot the wire length is 27 ft.

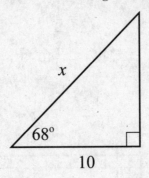

28. The area of a sector is found using $A_{sector} = \pi R^2 \dfrac{\theta}{360}$ where θ is the measure of the central angle in degrees and R is the radius. The central angle is equal in measure to the intercepted arc, which in this case is 72°. Substitute the $\theta = 72$ and $R = 10$ to find the area.

$$A_{sector} = \pi R^2 \frac{\theta}{360}$$

$$= \pi 10^2 \frac{72}{360}$$

$$= 100\pi$$

The area of the shaded sector is 20π.

29. Start with one pair of coordinates and look for one or more transformations that will map one point to another. Be sure to confirm your transformations work for the other pairs of points. Taking points $A(-2, 5)$ and $T(-5, -2)$, we see that a rotation of 90° clockwise about the origin would map A to T since the rotation transforms (x, y) to $(-y, x)$. The same transformation also maps C to U and B to T. A rotation of 90° counterclockwise about the origin maps $\triangle ABC$ to $\triangle STU$.

Note that there are numerous other correct answers, such as a rotation of 90° counterclockwise about point A followed by a translation 6 down and left 3.

30. Altitude $\overline{PQ}$ drawn to a hypotenuse $\overline{RT}$ forms two triangles similar to the original. From the given side lengths, we are working with $\triangle PRT$ and $\triangle QRP$. Sketch the triangles, matching up corresponding angles. Angles $\angle TPR$ and $\angle PQR$ are congruent right angles. The shared angle at R is also congruent in each. The similarity statement is $\triangle PRT \sim \triangle QRP$. Corresponding sides of similar triangles are proportional, so we can set up a proportion and solve for RQ.

$$\frac{RQ}{PR} = \frac{PR}{RT}$$

$$\frac{RQ}{15} = \frac{15}{17}$$

$$17RQ = 15 \cdot 15$$

$$RQ = \frac{225}{17}$$

$$= 13.23529$$

$RQ = 13.2$ rounded to the nearest tenth.

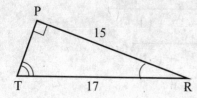 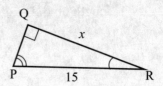

31. Use your compass to measure the radius $\overline{OA}$ by putting the point at O and the pencil at A. Keep the same compass opening and put the point at A and make an arc at B. Move the compass point to B and make an arc at C. Continue making arcs at D, E, and F. If you are accurate, a final arc should bring you back to point A. Connect points A, C, and E. $\triangle ACE$ is an equilateral triangle.

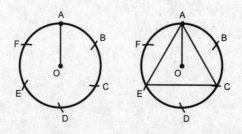

PART III

32. A trapezoid has at least one pair of parallel sides. From the graph, it appears that $\overline{AD}$ and $\overline{BC}$ may be parallel. We can check if they are parallel by calculating the slope of each.

$$\text{slope} = \frac{y_2 - y_1}{x_2 - x_1}$$

$$\text{slope of } \overline{AD} = \frac{0 - 6}{1 - (-1)} = -3$$

$$\text{slope of } \overline{BC} = \frac{-1 - 8}{6 - 3} = -3$$

The slopes of $\overline{AD}$ and $\overline{BC}$ are equal; therefore, they are parallel. *ABCD* is a trapezoid because it has a pair of parallel sides.

We can check if the trapezoid is isosceles by calculating the lengths of the diagonals $\overline{AC}$ and $\overline{BD}$.

$$\text{length} = \sqrt{(x_2 - x_1)^2 + (y_2 - y_1)^2}$$

$$\text{length of } \overline{BD} = \sqrt{(1 - 3)^2 + (0 - 8)^2}$$

$$= \sqrt{(-2)^2 + (-8)^2}$$

$$= \sqrt{4 + 64}$$

$$= \sqrt{68}$$

$$\text{length of } \overline{AC} = \sqrt{(6 - (-1))^2 + (-1 - 6)^2}$$

$$= \sqrt{(7)^2 + (-7)^2}$$

$$= \sqrt{49 + 49}$$

$$= \sqrt{98}$$

The lengths of $\overline{BD}$ and $\overline{AC}$ are not equal. The diagonals are not congruent so *ABCD* is not isosceles.

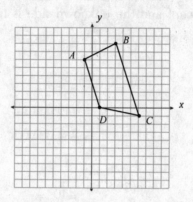

33. The volume of a cylinder is found using $V = \pi R^2 h$ where R is the radius and h is the height. The radius is half the diameter, or 3.25 ft. Since we want our answer in cubic feet, we need to convert the height to feet. 12 inches is equivalent to 1 ft, so the height equals 1 ft.

$$V = \pi R^2 h$$

$$= \pi (3.25)^2 (1)$$

$$= 33.183072 \text{ ft}^3$$

The pool is filled only $\frac{2}{3}$ full, so multiply the volume by $\frac{2}{3}$.

$$V_{water} = \frac{2}{3}(33.183072)$$

$$= 22.122048 \text{ ft}^3$$

The volume of water is 22 ft³ rounded the nearest cubic foot.

Find the number of gallons by multiplying the volume by the given conversion factor.

$$22 \text{ ft}^3 \cdot 7.48 \ \frac{\text{gallons}}{\text{ft}^3} = 164.56 \text{ gallons}$$

To the nearest gallon, the amount of water is 165 gallons.

34. The figure can be broken up into the two right triangles shown below. The elevation of the bottom of the blade is y, and the elevation of the top of the blade is z. Use two trig ratios to calculate y and z. In each triangle, the 440-ft distance is the adjacent and the unknown side is the opposite, so we use tangent ratios.

Calculate y:

$$\tan(30°) = \frac{y}{440}$$
$$y = 440 \tan(30°)$$
$$= 254.034118$$

Calculate z:

$$\tan(38.8°) = \frac{z}{440}$$
$$z = 440 \tan(38.8°)$$
$$= 353.769082$$

The blade length x is found by subtracting y from z.

$$x = z - y$$
$$= 353.769082 - 254.034118$$
$$= 99.734964$$

Rounded to the nearest foot the blade length is 100 ft.

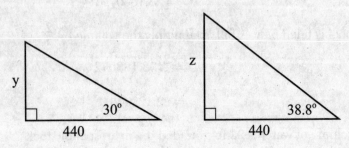

PART IV

35. The strategy for this proof is to show that $\triangle HEA$ is similar to $\triangle TAH$. Corresponding sides in these triangles are proportional, and the proportion leads to $TA \cdot HA = HE \cdot TH$ when cross multiplied. In order to prove the triangles similar, we start by proving $MHTA$ is a parallelogram so we can make use of its parallel sides.

Statement	Reason
1. $\overline{HM} \cong \overline{AT}$, $\overline{HT} \cong \overline{AM}$	1. Given
2. $MHTA$ is a parallelogram	2. A quadrilateral with opposite sides congruent is a parallelogram
3. $\overline{HT} \parallel \overline{MA}$	3. Opposite sides of a parallelogram are parallel
4. $\angle THA \cong \angle EAH$	4. Alternate interior angles formed by parallel lines are congruent
5. $\overline{HE} \perp \overline{MEA}$, $\overline{HA} \perp \overline{AT}$	5. Given
6. $\angle HEA$ and $\angle TAH$ are right angles	6. Perpendicular lines form right angles
7. $\angle HEA \cong \angle TAH$	7. Right angles are congruent
8. $\triangle HEA \sim \triangle TAH$	8. AA
9. $\dfrac{TA}{HE} = \dfrac{TH}{HA}$	9. Corresponding sides in similar triangles are proportional
10. $TA \cdot HA = HE \cdot TH$	10. Cross products of a proportion are equal

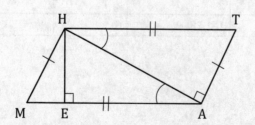

Topic	Question Numbers	Number of Points	Your Points	Your Percentage
1. Basic Angle and Segment Relationships				
2. Angle and Segment Relationships in Triangles and Polygons	21, 23	4		
3. Constructions	31	2		
4. Transformations	1, 4, 5, 29	8		
5. Triangle Congruence	8, 14, 25	6		
6. Lines, Segments, and Circles on the Coordinate Plane	7, 16, 19, 20, 26	10		
7. Similarity	11, 15, 30, 35	12		
8. Trigonometry	9, 22, 27, 34	10		
9. Parallelograms	12, 17, 24	6		
10. Coordinate Geometry Proofs	32	4		
11. Volume and Solids	3, 6, 10, 33	10		
12. Modeling	2	2		
13. Circles	13, 18, 28	6		

HOW TO CALCULATE YOUR GEOMETRY REGENTS EXAM SCORE

You will need to convert your raw score out of 85 points to a scaled score, which will be your official Geometry Regents grade. Add the total number of credits you scored on the exam to get your raw score. Partial credit is allowed on Parts II, III, and IV so you may need to estimate the number of credits for questions not completely correct. Your scaled score is found using the accompanying conversion chart. Find your raw score in the column labeled "Raw Score". Your final Geometry Regents Exam score is the corresponding number in the "Scaled Score" column. The scaled score is what will be reported to you by your school.

Regents Exam in Geometry—June 2019
Chart for Converting Total Test Raw Scores to
Final Exam Scores (Scale Scores)

Raw Score	Scale Score	Performance Level	Raw Score	Scale Score	Performance Level	Raw Score	Scale Score	Performance Level
80	100	5	53	80	4	26	61	2
79	99	5	52	79	3	25	60	2
78	98	5	51	79	3	24	58	2
77	97	5	50	78	3	23	57	2
76	96	5	49	78	3	22	55	2
75	95	5	48	77	3	21	54	1
74	94	5	47	77	3	20	52	1
73	93	5	46	76	3	19	51	1
72	92	5	45	76	3	18	49	1
71	91	5	44	75	3	17	47	1
70	90	5	43	75	3	16	45	1
69	90	5	42	74	3	15	43	1
68	89	5	41	74	3	14	41	1
67	88	5	40	73	3	13	39	1
66	87	5	39	73	3	12	36	1
65	87	5	38	72	3	11	34	1
64	86	5	37	71	3	10	32	1
63	86	5	36	71	3	9	29	1
62	85	5	35	70	3	8	26	1
61	84	4	34	69	3	7	23	1
60	84	4	33	68	3	6	20	1
59	83	4	32	67	3	5	17	1
58	82	4	31	66	3	4	14	1
57	82	4	30	65	3	3	11	1
56	81	4	29	64	2	2	7	1
55	81	4	28	63	2	1	4	1
54	80	4	27	62	2	0	0	1

August 2019 Exam
Geometry

HIGH SCHOOL MATH REFERENCE SHEET

1 inch = 2.54 centimeters	1 cup = 8 fluid ounces
1 meter = 39.37 inches	1 pint = 2 cups
1 mile = 5280 feet	1 quart = 2 pints
1 mile = 1760 yards	1 gallon = 4 quarts
1 mile = 1.609 kilometers	1 gallon = 3.785 liters
	1 liter = 0.264 gallon
1 kilometer = 0.62 mile	1 liter = 1000 cubic centimeters
1 pound = 16 ounces	
1 pound = 0.454 kilogram	
1 kilogram = 2.2 pounds	
1 ton = 2000 pounds	

Triangle	$A = \dfrac{1}{2}bh$
Parallelogram	$A = bh$
Circle	$A = \pi r^2$
Circle	$C = \pi d$ or $C = 2\pi r$

General Prism	$V = Bh$
Cylinder	$V = \pi r^2 h$
Sphere	$V = \frac{4}{3}\pi r^3$
Cone	$V = \frac{1}{3}\pi r^2 h$
Pyramid	$V = \frac{1}{3}Bh$
Pythagorean Theorem	$a^2 + b^2 = c^2$
Quadratic Formula	$x = \dfrac{-b \pm \sqrt{b^2 - 4ac}}{2a}$
Arithmetic Sequence	$a_n = a_1 + (n-1)d$
Geometric Sequence	$a_n = a_1 r^{n-1}$
Geometric Series	$S_n = \dfrac{a_1 - a_1 r^n}{1-r}$ where $r \neq 1$
Radians	$1 \text{ radian} = \dfrac{180}{\pi}$ degrees
Degrees	$1 \text{ degree} = \dfrac{\pi}{180}$ radians
Exponential Growth/Decay	$A = A_0 e^{k(t-t_0)} + B_0$

180 Minutes–35 Questions

PART I

Answer all 24 questions in this part. Each correct answer will receive 2 credits. No partial credit will be allowed. For each statement or question, write in the space provided the numeral preceding the word or expression that best completes the statement or answers the question.   [48 credits]

1 On the set of axes below, $\overline{AB}$ is dilated by a scale factor of $\frac{5}{2}$ centered at point P.

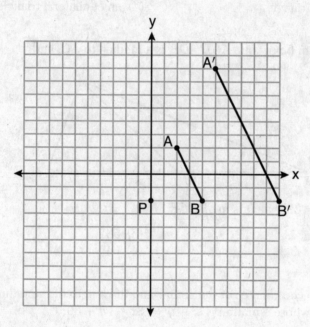

Which statement is always true?

(1) $\overline{PA} \cong \overline{AA'}$

(2) $\overline{AB} \parallel \overline{A'B'}$

(3) $AB = A'B'$

(4) $\frac{5}{2}(A'B') = AB$ 1 _____

2 The coordinates of the vertices of parallelogram $CDEH$ are $C(-5,5)$, $D(2,5)$, $E(-1,-1)$, and $H(-8,-1)$. What are the coordinates of P, the point of intersection of diagonals $\overline{CE}$ and $\overline{DH}$?

(1) $(-2,3)$

(2) $(-2,2)$

(3) $(-3,2)$

(4) $(-3,-2)$ 2 _____

3 The coordinates of the endpoints of $\overline{QS}$ are $Q(-9,8)$ and $S(9,-4)$.
   Point R is on $\overline{QS}$ such that $QR : RS$ is in the ratio of $1 : 2$. What are
   the coordinates of point R?

   (1) (0,2)   (3) (−3,4)
   (2) (3,0)   (4) (−6,6)   3 _____

4 If the altitudes of a triangle meet at one of the triangle's vertices,
   then the triangle is

   (1) a right triangle   (3) an obtuse triangle
   (2) an acute triangle   (4) an equilateral triangle   4 _____

5 In the diagram below of △ACD, $\overline{DB}$ is a median to $\overline{AC}$, and
   $AB \simeq DB$.

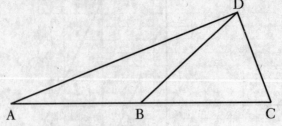

   If m∠DAB = 32°, what is m∠BDC?

   (1) 32°   (3) 58°
   (2) 52°   (4) 64°   5 _____

6 What are the coordinates of the center and the length of the radius
   of the circle whose equation is $x^2 + y^2 = 8x - 6y + 39$?

   (1) center (−4,3) and radius 64
   (2) center (4,−3) and radius 64
   (3) center (−4,3) and radius 8
   (4) center (4,−3) and radius 8   6 _____

7 In the diagram below of parallelogram $ABCD$, $\overline{AFGB}$, $\overline{CF}$ bisects $\angle DCB$, $\overline{DG}$ bisects $\angle ADC$, and $\overline{CF}$ and $\overline{DG}$ intersect at E.

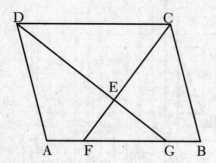

If $m\angle B = 75°$, then the measure of $\angle EFA$ is

(1) 142.5°

(3) 52.5°

(2) 127.5°

(4) 37.5°   7____

8 What is an equation of a line that is perpendicular to the line whose equation is $2y + 3x = 1$?

(1) $y = \frac{2}{3}x + \frac{5}{2}$

(3) $y = -\frac{2}{3}x + 1$

(2) $y = \frac{3}{2}x + 2$

(4) $y = -\frac{3}{2}x + \frac{1}{2}$ 8____

9 Triangles *ABC* and *RST* are graphed on the set of axes below.

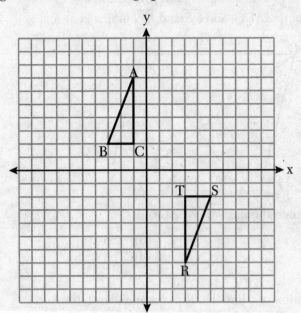

Which sequence of rigid motions will prove $\triangle ABC \cong \triangle RST$?

(1) · a line reflection over $y = x$

(2) a rotation of 180° centered at (1,0)

(3) a line reflection over the *x*-axis followed by a translation of 6 units right

(4) a line reflection over the *x*-axis followed by a line reflection over $y = x$ 9 _____

10 If the line represented by $y = -\dfrac{1}{4}x - 2$ is dilated by a scale factor of 4 centered at the origin, which statement about the image is true?

(1) The slope is $-\dfrac{1}{4}$ and the *y*-intercept is -8.

(2) The slope is $-\dfrac{1}{4}$ and the *y*-intercept is -2.

(3) The slope is -1 and the *y*-intercept is -8.

(4) The slope is -1 and the *y*-intercept is -2.   10 _____

11 Square *MATH* has a side length of 7 inches. Which three-dimensional object will be formed by continuously rotating square *MATH* around side $\overline{AT}$?

   (1) a right cone with a base diameter of 7 inches

   (2) a right cylinder with a diameter of 7 inches

   (3) a right cone with a base radius of 7 inches

   (4) a right cylinder with a radius of 7 inches   11 _____

12 Circle *O* with a radius of 9 is drawn below. The measure of central angle *AOC* is 120°.

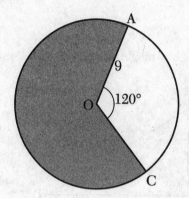

What is the area of the shaded sector of circle *O*?

   (1) 6π (3) 27π

   (2) 12π (4) 54π 12 _____

13 In quadrilateral *QRST*, diagonals $\overline{QS}$ and $\overline{RT}$ intersect at *M*. Which statement would always prove quadrilateral *QRST* is a parallelogram?

   (1) $\angle TQR$ and $\angle QRS$ are supplementary.

   (2) $\overline{QM} \cong \overline{SM}$ and $\overline{QT} \cong \overline{RS}$

   (3) $\overline{QR} \cong \overline{TS}$ and $\overline{QT} \cong \overline{RS}$

   (4) $\overline{QR} \cong \overline{TS}$ and $\overline{QT} \parallel \overline{RS}$ 13 _____

14 A standard-size golf ball has a diameter of 1.680 inches. The material used to make the golf ball weighs 0.6523 ounce per cubic inch. What is the weight, to the *nearest hundredth of an ounce*, of one golf ball?

(1) 1.10   (3) 2.48

(2) 1.62   (4) 3.81   14 _____

15 Chelsea is sitting 8 feet from the foot of a tree. From where she is sitting, the angle of elevation of her line of sight to the top of the tree is 36°. If her line of sight starts 1.5 feet above ground, how tall is the tree, to the *nearest foot*?

(1) 8   (3) 6

(2) 7   (4) 4   15 _____

16 In the diagram below of right triangle ABC, altitude $\overline{CD}$ intersects hypotenuse $\overline{AB}$ at D.

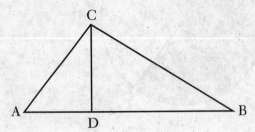

Which equation is always true?

(1) $\dfrac{AD}{AC} = \dfrac{CD}{BC}$ (3) $\dfrac{AC}{CD} = \dfrac{BC}{CD}$

(2) $\dfrac{AD}{CD} = \dfrac{BD}{CD}$ (4) $\dfrac{AD}{AC} = \dfrac{AC}{BD}$ 16 _____

17 A countertop for a kitchen is modeled with the dimensions shown below. An 18-inch by 21-inch rectangle will be removed for the installation of the sink.

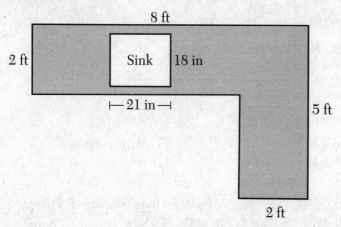

What is the area of the top of the installed countertop, to the *nearest square foot*?

(1) 26

(2) 23

(3) 22

(4) 19

17 _____

18 In the diagram below, $\overline{BC}$ connects points B and C on the congruent sides of isosceles triangle ADE, such that $\triangle ABC$ is isosceles with vertex angle A.

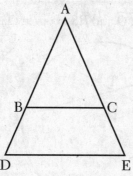

If $AB = 10$, $BD = 5$, and $DE = 12$, what is the length of $\overline{BC}$?

(1) 6

(2) 7

(3) 8

(4) 9

18 _____

19 In $\triangle ABC$ below, angle C is a right angle.

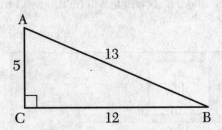

Which statement must be true?

(1) $\sin A = \cos B$ (3) $\sin B = \tan A$
(2) $\sin A = \tan B$ (4) $\sin B = \cos B$ 19 _____

20 In right triangle RST, altitude $\overline{TV}$ is drawn to hypotenuse $\overline{RS}$. If $RV = 12$ and $RT = 18$, what is the length of $\overline{SV}$?

(1) $6\sqrt{5}$ (3) $6\sqrt{6}$
(2) 15   (4) 27   20 _____

21 What is the volume, in cubic centimeters, of a right square pyramid with base edges that are 64 cm long and a slant height of 40 cm?

(1) 8192.0   (3) 32,768.0
(2) 13,653.$\overline{3}$ (4) 54,613.$\overline{3}$ 21 _____

22 In the diagram below, chords $\overline{PQ}$ and $\overline{RS}$ of circle O intersect at T.

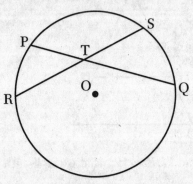

Which relationship must always be true?

(1) $RT = TQ$ (3) $RT + TS = PT + TQ$
(2) $RT = TS$ (4) $RT \times TS = PT \times TQ$ 22 _____

23 A rhombus is graphed on the set of axes below.

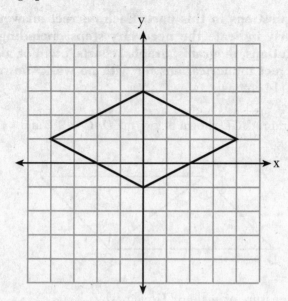

Which transformation would carry the rhombus onto itself?

(1) 180° rotation counterclockwise about the origin

(2) reflection over the line $y = \frac{1}{2}x + 1$

(3) reflection over the line $y = 0$

(4) reflection over the line $x = 0$ 23 _____

24 A 15-foot ladder leans against a wall and makes an angle of 65°
with the ground. What is the horizontal distance from the wall to
the base of the ladder, to the *nearest tenth of a foot*?

(1) 6.3   (3) 12.9

(2) 7.0   (4) 13.6   24 _____

PART II

Answer all 7 questions in this part. Each correct answer will receive 2 credits. Clearly indicate the necessary steps, including appropriate formula substitutions, diagrams, graphs, charts, etc. For all questions in this part, a correct numerical answer with no work shown will receive only 1 credit. [14 credits]

25 In parallelogram *ABCD* shown below, m∠*DAC* = 98° and
   m∠*ACD* = 36°.

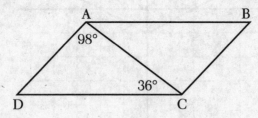

What is the measure of angle *B*? Explain why.

26 An airplane took off at a constant angle of elevation. After the plane traveled for 25 miles, it reached an altitude of 5 miles, as modeled below.

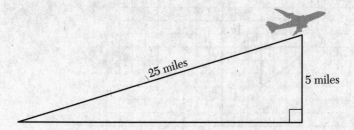

To the *nearest tenth of a degree*, what was the angle of elevation?

27 On the set of axes below, $\triangle ABC \cong \triangle DEF$.

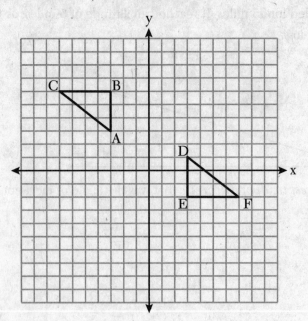

Describe a sequence of rigid motions that maps $\triangle ABC$ onto $\triangle DEF$.

28 The vertices of $\triangle ABC$ have coordinates $A(-2,-1)$, $B(10,-1)$, and $C(4,4)$. Determine and state the area of $\triangle ABC$. [The use of the set of axes below is optional.]

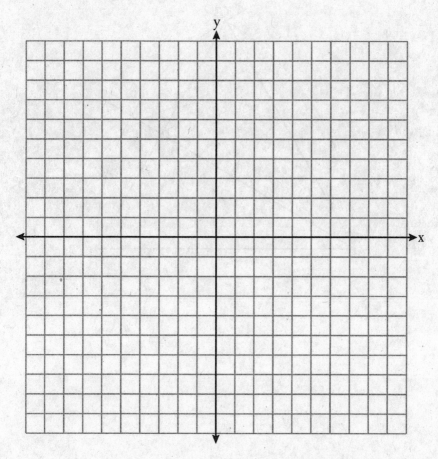

29 Using the construction below, state the degree measure of ∠*CAD*. Explain why.

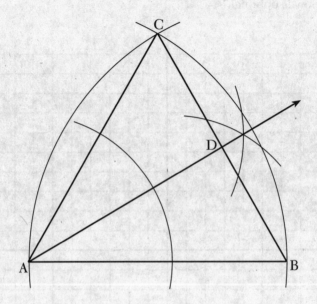

30 In the diagram below of circle K, secant $\overline{PLKE}$ and tangent $\overline{PZ}$ are drawn from external point P.

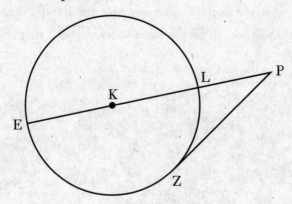

If $m\widehat{LZ} = 56°$, determine and state the degree measure of angle P.

31 A large water basin is in the shape of a right cylinder. The inside of the basin has a diameter of $8\frac{1}{4}$ feet and a height of 3 feet. Determine and state, to the *nearest cubic foot*, the number of cubic feet of water that it will take to fill the basin to a level of $\frac{1}{2}$ foot from the top.

PART III

Answer all 3 questions in this part. Each correct answer will receive 4 credits. Clearly indicate the necessary steps, including appropriate formula substitutions, diagrams, graphs, charts, etc. For all questions in this part, a correct numerical answer with no work shown will receive only 1 credit.　[12 credits]

32　Triangle *ABC* is shown below. Using a compass and straightedge, construct the dilation of △*ABC* centered at *B* with a scale factor of 2. [Leave all construction marks.]

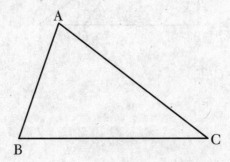

Is the image of △*ABC* similar to the original triangle? Explain why.

33 In the diagram below, $\triangle ABE \cong \triangle CBD$.

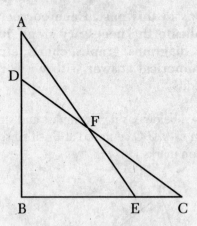

Prove: $\triangle AFD \cong \triangle CFE$

34 A cargo trailer, pictured below, can be modeled by a rectangular prism and a triangular prism. Inside the trailer, the rectangular prism measures 6 feet wide and 10 feet long. The walls that form the triangular prism each measure 4 feet wide inside the trailer. The diagram below is of the floor, showing the inside measurements of the trailer.

Cargo Trailer

Cargo Trailer Floor

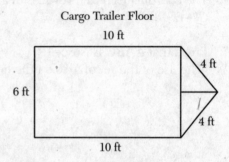

If the inside height of the trailer is 6.5 feet, what is the total volume of the inside of the trailer, to the *nearest cubic foot*?

PART IV

Answer the question in this part. A correct answer will receive 6 credits. Clearly indicate the necessary steps, including appropriate formula substitutions, diagrams, graphs, charts, etc. A correct numerical answer with no work shown will receive only 1 credit.   [6 credits]

35 The coordinates of the vertices of $\triangle ABC$ are $A(1,2)$, $B(-5,3)$, and $C(-6,-3)$.

Prove that $\triangle ABC$ is isosceles.
[The use of the set of axes on the next page is optional.]

State the coordinates of point D such that quadrilateral $ABCD$ is a square.

Question 35 is continued on the next page.

Question 35 continued

Prove that your quadrilateral *ABCD* is a square.
[The use of the set of axes below is optional.]

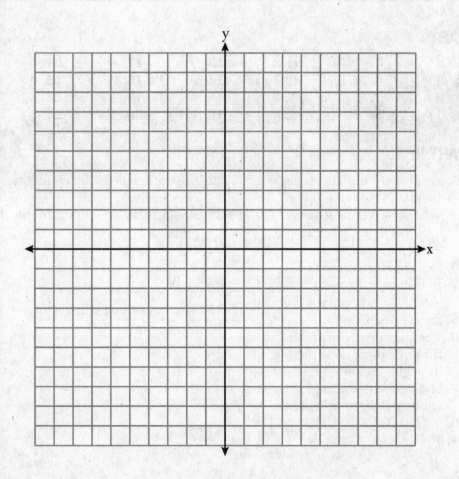

Answers August 2019
Geometry

Answer Key

PART I

1. 2	**5.** 3	**9.** 2	**13.** 3	**17.** 4	**21.** 3
2. 3	**6.** 4	**10.** 1	**14.** 2	**18.** 3	**22.** 4
3. 3	**7.** 2	**11.** 4	**15.** 2	**19.** 1	**23.** 4
4. 1	**8.** 1	**12.** 4	**16.** 1	**20.** 2	**24.** 1

PART II

25. m$\angle D = 46°$ from the triangle angle sum theorem, and m$\angle B =$ m$\angle D = 46°$ because opposite angles of a parallelogram are congruent.

26. angle of elevation $= 11.5°$

27. Reflect over the x-axis, reflect over the y-axis, and translate 4 up.

28. area of $\triangle ABC = 30$

29. The triangle is constructed using the equilateral construction, so its angles are $60°$. $\angle BAC$ is bisected with the angle bisector construction, so m$\angle CAD = 30°$.

30. m$\angle P = 34°$

31. $134\,\text{ft}^3$

PART III

32. See the detailed solution for the construction.

33. See the detailed solution for the proof.

34. $442\,\text{ft}^3$

PART IV

35. See the detailed solution for the proof.

In **PARTS II–IV**, you are required to show how you arrived at your answers. For sample methods of solutions, see the *Answers Explained* section.

Answers Explained

PART I

1. The dilation of a line segment is always parallel to the original segment, so $\overline{AB} \parallel \overline{A'B'}$. Examining the other choices, we see that choice (1) is incorrect because $\overline{PA} \cong \overline{AA'}$ implies a $2:1$ ratio instead of the $5:2$ ratio. Choices (3) and (4) are incorrect because $A'B' = \dfrac{5}{2}AB$.

   The correct choice is **(2)**.

2. The diagonals of a parallelogram bisect each other, so the point of intersection must be the midpoint of the diagonals. Using the coordinates $(-5, 5)$ and $(-1, -1)$ for diagonal $\overline{CE}$, apply the midpoint formula.

$$\text{midpoint} = \left(\frac{x_1 + x_2}{2}, \frac{y_1 + y_2}{2} \right)$$

$$= \left(\frac{-5 + (-1)}{2}, \frac{5 + (-1)}{2} \right)$$

$$= (-3, 2)$$

   The correct choice is **(3)**. Note that you would get the same result using diagonal $\overline{DH}$.

3. The coordinates of the point that divide a segment in a specified ratio can be calculated using the following formula:

$$\text{ratio} = \frac{x - x_1}{x_2 - x} \qquad \text{ratio} = \frac{y - y_1}{y_2 - y}$$

   where the coordinates of $Q(-9, 8)$ are (x_1, y_1) and the coordinates of $S(9, -4)$ are (x_2, y_2).

$$\frac{1}{2} = \frac{x - (-9)}{9 - x} \qquad\qquad \frac{1}{2} = \frac{y - 8}{-4 - y}$$

$$2x + 18 = 9 - x \qquad\qquad 2y - 16 = -4 - y$$

$$3x = -9 \qquad\qquad\qquad 3y = 12$$

$$x = -3 \qquad\qquad\qquad\quad y = 4$$

   The coordinates of R are $(-3, 4)$.

   The correct choice is **(3)**.

4. Altitudes of a triangle always form a right angle with the opposite side. If the altitudes meet at a vertex, then the triangle must be a right triangle.

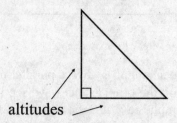

altitudes

The correct choice is (**1**).

5. Since $\overline{AB} \cong \overline{BD}$, we know $\triangle ABD$ is isosceles and $\angle DAB \cong \angle ADB$. Apply the angle sum theorem in $\triangle ABD$ to find m$\angle ABD$.

$$m\angle ABD + m\angle DAB + m\angle ADB = 180°$$
$$m\angle ABD + 32° + 32° = 180°$$
$$m\angle ABD = 116°$$

Next, we see that $\angle ABD$ and $\angle CBD$ are a linear pair and sum to 180°.

$$m\angle ABD + m\angle CBD = 180°$$
$$116° + m\angle CBD = 180°$$
$$m\angle CBD = 64°$$

A median divides a side of a triangle into two congruent segments, so $\overline{AB} \cong \overline{BC}$. Combining this with $\overline{AB} \cong \overline{BD}$, we can conclude that $\overline{BC} \cong \overline{BD}$. We now know that $\triangle BDC$ is isosceles with $\angle BDC \cong \angle BCD$. Apply the triangle sum theorem to find m$\angle BDC$.

$$m\angle BDC + m\angle BCD + m\angle CBD = 180°$$
$$2\, m\angle BDC + 64° = 180°$$
$$2\, m\angle BDC = 116°$$
$$m\angle BDC = 58°$$

The correct choice is (**3**).

6. Use the completing the square procedure to convert the equation to center-radius form, $(x - h)^2 + (y - k^2) = R^2$. The center has coordinates (h, k) and the radius is R. Start by rearranging the equation so all the x and y terms are grouped together on the left.

$$x^2 + y^2 = 8x - 6y + 39$$
$$x^2 - 8x + y^2 + 6y = 39$$

Now find the constant terms we need to complete the square:

Constant for the x-terms $= (\frac{1}{2} \cdot \text{coefficient of } x)^2$

$$= \left(\frac{1}{2} \cdot (-8)\right)^2$$
$$= 16$$

Constant for the y-terms $= (\frac{1}{2} \cdot \text{coefficient of } y)^2$

$$= \left(\frac{1}{2} \cdot 6\right)^2$$
$$= 9$$

Add the required constant to each side of the equation and factor.

$$x^2 - 8x + 16 + y^2 + 6y + 9 = 39 + 16 + 9$$
$$(x - 4)^2 + (y + 3)^2 = 64$$

The values of h and k are 4 and 3. Be careful with the signs because h and k are the values subtracted from x and y. The center is located at $(4, -3)$. Since $R^2 = 64$, the radius is equal to $\sqrt{64}$, or 8.

The correct choice is **(4)**.

7. Consecutive angles of a parallelogram are supplementary, so we can find $m\angle BCD$.

$$m\angle BCD + m\angle B = 180°$$
$$m\angle BCD + 75° = 180°$$
$$m\angle BCD = 105°$$

Next, use angle bisector $\overline{CF}$ to find m$\angle DCF$.

$$m\angle DCF = \frac{1}{2}\, m\angle BCD$$
$$= \frac{1}{2}(105°)$$
$$= 52.5°$$

$\angle DCF$ and $\angle CFB$ are congruent alternate interior angles, so m$\angle CFG$ is also equal to 52.5°. Finally, $\angle EFA$ and $\angle CFG$ are a linear pair that sum to 180°.

$$m\angle EFA + m\angle CFG = 180°$$
$$m\angle EFA + 52.5° = 180°$$
$$m\angle EFA = 127.5°$$

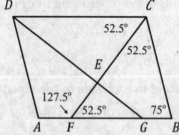

The correct choice is (**2**).

8. First, identify the slope by rewriting the equation in $y = mx + b$ form.

$$2y + 3x = 1$$
$$2y = -3x + 1$$
$$y = -\frac{3}{2}x + \frac{1}{2}$$

The slope of the given line is $-\frac{3}{2}$. Slopes of perpendicular lines are negative reciprocals, which we can find by changing the sign and flipping the fraction. The negative reciprocal of $-\frac{3}{2}$ is $\frac{2}{3}$. The only choice with slope of $\frac{2}{3}$ is $y = \frac{2}{3}x + \frac{5}{2}$.

The correct choice is (**1**).

9.

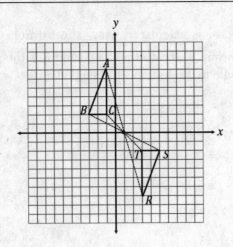

Two figures are congruent if a sequence of rigid motions maps one figure onto another. To check if a 180° rotation was used, sketch segments from each vertex of $\triangle ABC$ to the corresponding vertex of $\triangle RST$ as shown in the figure. All three segments are concurrent at $(1, 0)$, so this point is the center of a 180° rotation that maps $\triangle ABC$ to $\triangle RST$.

The correct choice is (**2**).

10. When a line is dilated about the origin, the slope remains the same, but the y-intercept is multiplied by the scale factor. The slope remains $-\frac{1}{4}$. The y-intercept transforms to $4 \cdot (-2)$, or -8.

The correct choice is (**1**).

11. A square rotated about any of its sides always sweeps out a cylinder. The side length of the square is the height, as well as the radius of the cylinder. The solid is a cylinder with a radius of 7 inches.

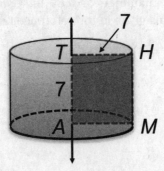

The correct choice is (**4**).

12. The area of a sector is calculated using the formula $A_{sector} = \pi R^2 \dfrac{\theta}{360}$, where θ is the measure of the central angle and R is the radius. First, calculate the central angle that intercepts the shaded arc.

$$\theta = 360° - 120°$$
$$= 240°$$
$$= A_{sector} = \pi R^2 \dfrac{\theta}{360}$$
$$= \pi \cdot 9^2 \cdot \dfrac{240}{360}$$
$$= 54\pi$$

The correct choice is (**4**).

13. One possible property that would prove a quadrilateral is a parallelogram is congruent opposite sides. For quadrilateral $QRST$, congruent opposite sides would be $\overline{QR} \cong \overline{TS}$ and $\overline{QT} \cong \overline{RS}$.

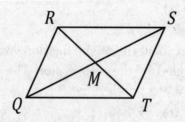

The correct choice is (**3**).

14. The golf ball can be modeled as a sphere with a diameter of 1.680 in. Its radius is one half the diameter, or 0.84 in. Calculate the volume of the sphere using the formula given in the reference table.

$$V = \dfrac{4}{3}\pi R^3$$
$$= \dfrac{4}{3}\pi (0.84)^3$$
$$= 2.482713 \text{ in}^3$$

The weight of an object is equal to the product of its volume and density.

$$\text{weight} = \text{volume} \cdot \text{density}$$

$$= 2.482713 \text{ in}^3 \left(0.6523 \frac{\text{ounces}}{\text{in}^3} \right)$$

$$= 1.62 \text{ ounces}$$

The correct choice is **(2)**.

15. The tree and Chelsea's line of sight form a right triangle as shown in the figure. Relative to the 36° angle of elevation, the 8-foot distance is the adjacent and the observed tree height is the opposite. A tangent ratio can be used to find the unknown height.

$$\tan(36°) = \frac{\text{opposite}}{\text{adjacent}}$$

$$\tan(36°) = \frac{x}{8}$$

$$x = 8 \tan(36°)$$

$$= 5.8123 \text{ ft}$$

Add the 1.5-ft distance from the ground to Chelsea's line of sight to get the total height.

$$\text{total height} = 5.8123 + 1.5$$

$$= 7.3123 \text{ ft}$$

To the nearest foot the height of the tree is 7 feet.

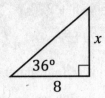

The correct choice is **(2)**.

16. When an altitude is drawn to the hypotenuse of a right triangle, three similar triangles are formed. Sketch the three triangles separately and label the congruent angles. We know that corresponding sides of similar triangles are congruent, so we can write the following proportion:

$$\frac{AD}{AC} = \frac{CD}{BC}$$

$\overline{AD}$ and $\overline{CD}$ are corresponding sides of the small and medium triangles. $\overline{AC}$ and $\overline{BC}$ are also corresponding sides of the same two triangles. None of the other choices match up corresponding sides of the same two triangles.

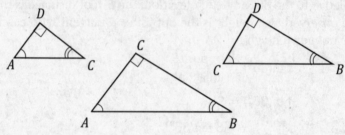

The correct choice is (**1**).

17. The countertop can be broken up into three rectangles—a solid 2 ft by 8 ft rectangle, a solid 2 ft by 3 ft rectangle, and a 21 in × 18 in cutout. Calculate the area of two solid rectangles.

$$\text{Area}_{\text{solid}} = \text{length}_1 \cdot \text{width}_1 + \text{length}_2 \cdot \text{width}_2$$

$$= 2 \cdot 8 + 2 \cdot 3$$

$$= 22 \text{ ft}^2$$

Before finding its area, the sink cutout dimensions need to be converted to feet by multiplying by the conversion factor $\dfrac{1 \text{ ft}}{12 \text{ in}}$.

Sink width $= 21 \text{ in} \cdot \dfrac{1 \text{ ft}}{12 \text{ in}}$ Sink length $= 18 \text{ in} \cdot \dfrac{1 \text{ ft}}{12 \text{ in}}$

$= 1.75 \text{ ft}$ $= 1.5 \text{ ft}$

$\text{Area}_{\text{sink}} = (1.75)(1.5)$

$= 2.625 \text{ ft}^2$

Now subtract the sink cutout for the sink from the solid area.

$$\text{Area}_{\text{countertop}} = 22 - 2.625$$

$$= 19.375 \text{ ft}^2$$

Rounded to the nearest square foot the area is 19 ft^2.

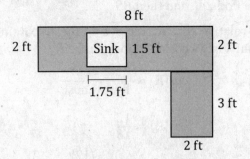

The correct choice is **(4)**.

18. Using the definition of isosceles triangles, we know $\overline{AB} \cong \overline{AC}$ and $\overline{AD} \cong \overline{AE}$. Dividing these congruence statements, we get $\frac{AB}{AD} \cong \frac{AC}{AE}$. The shared angle $\angle A$ is congruent in each triangle as well, so $\triangle BAC$ is similar to $\triangle DAE$ by the SAS postulate. Apply the theorem that corresponding sides in similar triangles are proportional.

$$\frac{BA}{DA} = \frac{BC}{DE}$$

$$\frac{10}{10+5} = \frac{BC}{12}$$

$$\frac{10}{15} = \frac{BC}{12}$$

$$15BC = 120$$

$$BC = 8$$

The correct choice is **(3)**.

19. Given two complementary angles A and B, the co-function relationship states that $\sin(A) = \cos(B)$. In right triangle ABC, angle C is a right triangle, so $\angle A$ and $\angle B$ are complementary. Therefore, $\sin(A) = \cos(B)$.

The correct choice is **(1)**.

20. There are several ways to approach this problem. One method is to apply trigonometry to find $\angle R$, and then RS.

Using $\triangle TVR$, relative to $\angle R$ side $\overline{RT}$ is the hypotenuse and $\overline{RV}$ is the adjacent. We can apply a cosine ratio to find $\angle R$.

$$\cos(R) = \frac{\text{adjacent}}{\text{hypotenuse}}$$

$$\cos(R) = \frac{12}{18}$$

$$R = \cos^{-1}\left(\frac{12}{18}\right)$$

$$R = 48.189685°$$

Now use $\triangle TRS$ to find RS. Relative to $\angle R$ side $\overline{RS}$ is the hypotenuse and side $\overline{TR}$ is the adjacent. Use a cosine ratio to find $\overline{RS}$.

$$\cos(R) = \frac{\text{adjacent}}{\text{hypotenuse}}$$

$$\cos(48.189685) = \frac{18}{RS}$$

$$RS = \frac{18}{\cos(48.189685)}$$

$$= 27$$

From the figure,

$$SV = RS - 12$$

$$= 27 - 12$$

$$= 15$$

The correct choice is **(2)**.

21. The height, h, of the pyramid forms a right triangle with the base and the 40 cm slant height. Use the Pythagorean theorem to find the height h. Be sure to use half of the side length for the bottom of the right triangle.

$$a^2 + b^2 = c^2$$
$$h^2 + 32^2 = 40^2$$
$$h^2 + 1024 = 1600$$
$$h^2 = 576$$
$$h = 24$$

Now find the volume of the pyramid using the volume formula from the reference table. The area of the base, B, is the area of the square base.

$$V = \frac{1}{3}Bh$$
$$= \frac{1}{3}\left(64^2\right)24$$
$$= 32{,}768$$

The volume of the pyramid is 32,768 cm^3.

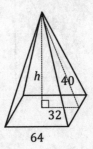

The correct choice is **(3)**.

22. When two chords intersect within a circle, the products of the parts of the chords are equal. In this case $RT \cdot TS = PT \cdot TQ$.

The correct choice is **(4)**.

23. The rhombus is symmetric about the y-axis, so a reflection over that axis will map the rhombus onto itself. The equation of the y-axis is $x = 0$.

The correct choice is (4).

24. The ladder, wall, and ground form a right triangle. We can use trigonometry to calculate the distance from the ladder to the wall. Relative to the 65° angle, the ladder is the hypotenuse and the distance to the wall is the adjacent. This indicates a cosine ratio.

$$\cos(65°) = \frac{\text{adjacent}}{\text{hypotenuse}}$$

$$\cos(65°) = \frac{x}{15}$$

$$x = 15\cos(65°)$$

$$= 6.33927$$

Rounded to the nearest foot, the distance to the wall is 6.3 ft.

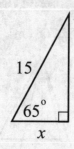

The correct choice is (1).

PART II

25. Two of the three angles are given in △ADC. Apply the angle sum theorem to find ∠D.

$$m\angle D + 36° + 98° = 180°$$

$$m\angle D + 134° = 180°$$

$$m\angle D = 46°$$

Opposite angles of a parallelogram are congruent, so m∠B = m∠D. Therefore, m∠B = 46°.

26. The angle of elevation is the angle formed between the ground and the 25-mile flight path. Relative to the angle of elevation, the 25-mile flight path is the hypotenuse and the 5-mile altitude is the opposite. This indicates using a sine ratio. Set up the correct ratio and then apply the inverse sine function to find the angle.

$$\sin(\theta) = \frac{5}{25}$$

$$\theta = \sin^{-1}\left(\frac{5}{25}\right)$$

$$= 11.53695°$$

Rounded to the nearest tenth of a degree, the angle of elevation is 11.5°.

27. There are many possible sequences of transformations that will map △ABC to △DEF. One possible sequence is two reflections followed by a translation. First, reflect △ABC over the x-axis to triangle I. Then, reflect triangle I over the y-axis to triangle II. Finally, translate triangle II up 4 to △DEF.

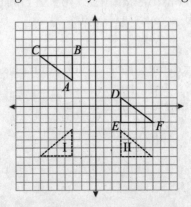

28. After graphing the triangle, sketch altitude $\overline{CD}$ to side $\overline{AB}$. We can calculate their lengths by counting grid units since they are vertical and horizontal segments. We find that $AB = 12$ and $CD = 5$. Now apply the formula for the area of a triangle.

$$A = \frac{1}{2}bh$$
$$= \frac{1}{2}(12)(5)$$
$$= 30$$

The area of $\triangle ABC$ is 30.

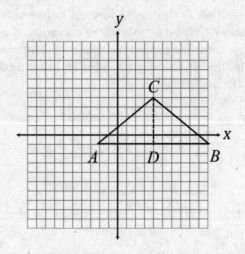

29. $\triangle ABC$ is constructed using the equilateral triangle construction. Since all angles of an equilateral triangle measure 60°, we know that m$\angle CAB = 60°$. The angle bisector construction was used to construct $\overrightarrow{AD}$. Bisector $\overrightarrow{AD}$ divides $\angle CAB$ in half, so m$\angle CAD = 30°$.

30. First find m$\widehat{EZ}$ using diameter $\overline{LE}$, which intercepts a 180° semicircle.

$$m\widehat{EZ} + m\widehat{LZ} = 180°$$
$$m\widehat{EZ} + 56° = 180°$$
$$m\widehat{EZ} = 124°$$

When a tangent and secant intersect, the angle formed is equal to one-half the difference of the intercepted arcs.

$$m\angle P = \frac{1}{2}\left(m\widehat{EZ} - m\widehat{LZ}\right)$$

$$= \frac{1}{2}(124° - 56°)$$

$$= 34°$$

31. Calculate the height of the water in the cylinder by subtracting $\frac{1}{2}$ ft from the height of the basin.

$$h_{\text{water}} = 3 - \frac{1}{2}$$

$$= 2.5 \text{ ft}$$

The radius of the cylinder is half the diameter, or 4.125 ft.

Next, calculate the volume of the water using the volume formula for a cylinder.

$$V = \pi R^2 h$$

$$= \pi(4.125^2)(2.5)$$

$$= 133.64041 \text{ ft}^3$$

To the nearest cubic foot, the volume of water is 134 ft^3.

PART III

32. (a) With the compass point at B, open the compass so the pencil is at A and make an arc to show that you measured that distance.

(b) Move the point to A and, with the same opening, make a second arc.

(c) Extend $\overline{BA}$ to the second arc. Label the intersection point E.

(d) Extend $\overline{BC}$ to F by repeating steps (a)–(c).

$\triangle EBF$ is a dilation of $\triangle ABC$ by a scale factor of 2 centered at B. The image is similar to the original because dilations are a similarity transformation. They increase length without changing any of the angle measures.

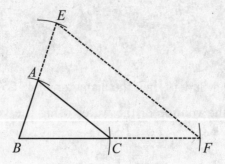

33. Prove the two triangles congruent by AAS. We will use a pair of congruent angles from the given congruent triangles, a pair of vertical angles, and sides $\overline{AD}$ and $\overline{CE}$. The sides are shown to be congruent by subtracting two pairs of congruent sides from the given triangles.

Statement	Reason
1. $\triangle ABE \cong \triangle CBD$	1. Given
2. $\angle A \cong \angle C$	2. CPCTC
3. $\angle AFD \cong \angle CFE$	3. Vertical angles are congruent
4. $\overline{AB} \cong \overline{CB}$	4. CPCTC
5. $\overline{BD} \cong \overline{BE}$	5. CPCTC
6. $\overline{AB} - \overline{BD} \cong \overline{CB} - \overline{BE}$	6. Subtraction Postulate
7. $\overline{AD} \cong \overline{CE}$	7. Partition Postulate
8. $\triangle AFD \cong \triangle CFE$	8. AAS

34. The volume of a prism is calculated using $V = Bh$, where B is the area of the base and h is the height of the prism. First, calculate the volume of the rectangular prism. The base is a 10 ft by 6 ft rectangle and the height is 6.5 ft.

$$B = 10 \cdot 6$$
$$= 60 \text{ ft}^2$$

$$V = Bh$$
$$= (60)(6.5)$$
$$= 390 \text{ ft}^3$$

The base of the triangular prism is an isosceles triangle with 4-ft legs and a base of 6 ft. To find its area we need to calculate the altitude, a, of the triangle. The altitude forms two right triangles as shown in the figure, and we can use the Pythagorean theorem to calculate a.

$$a^2 + b^2 = c^2$$
$$a^2 + 3^2 = 4^2$$
$$a^2 = 7$$
$$a = \sqrt{7}$$

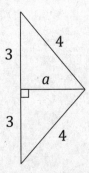

The area of the triangle base can now be calculated.

$$B = \frac{1}{2}(6)\left(\sqrt{7}\right)$$
$$= 3\sqrt{7}$$

Now calculate the volume of the triangular prism.

$$V = Bh$$
$$= \left(3\sqrt{7}\right)(6.5)$$

$$= 51.592 \text{ ft}^2$$

Finally combine the two volumes.

$$V_{\text{total}} = 390 \text{ ft}^3 + 51.592 \text{ ft}^3$$

$$441.592 \text{ ft}^3$$

Rounded to the nearest cubic foot the volume is 442 ft^3.

PART IV

35. An isosceles triangle has two congruent sides, so the proof requires us to show the lengths of two sides are equal. $\overline{AB}$ and $\overline{BC}$ appear to be the congruent sides, so start with those.

Length of AB:

$$d = \sqrt{(x_1 - x_2)^2 + (y_1 - y_2)^2}$$
$$= \sqrt{(1 - (-5))^2 + (2 - 3)^2}$$
$$= \sqrt{37}$$

Length of BC:

$$d = \sqrt{(x_1 - x_2)^2 + (y_1 - y_2)^2}$$
$$= \sqrt{(-5 - (-6))^2 + (3 - (-3))^2}$$
$$= \sqrt{37}$$

$\overline{AB}$ and $\overline{BC}$ have the same length, so they are congruent and $\triangle ABC$ is isosceles.

To find point D that completes the square, count the horizontal and vertical translations from point B to point A. Point A is 6 units to the right and 1 unit down from point B. Count those same distances from point C to locate point D. Starting from C $(-6, -3)$, 6 units right and 1 unit down puts point D at $(0, -4)$.

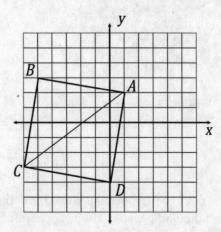

We need to demonstrate one parallelogram property, one rhombus property, and one rectangle property. One possible approach is to show:

- opposite sides are congruent (parallelogram)
- consecutive sides are congruent (rhombus)
- consecutive sides form a right angle (rectangle)

Since we already have the lengths of $\overline{AB}$ and $\overline{BC}$, we just need $\overline{CD}$ and $\overline{AD}$.

Length of CD:

$$d = \sqrt{\left(x_1 - x_2\right)^2 + \left(y_1 - y_2\right)^2}$$
$$= \sqrt{(-6 - 0)^2 + (-3 - (-4))^2}$$
$$= \sqrt{37}$$

Length of AD:

$$d = \sqrt{\left(x_1 - x_2\right)^2 + \left(y_1 - y_2\right)^2}$$
$$= \sqrt{(1 - 0)^2 + (2 - (-4))^2}$$
$$= \sqrt{37}$$

Now calculate the slopes of $\overline{AB}$ and $\overline{BC}$.

Slope of AB:

$$\text{slope} = \frac{y_2 - y_1}{x_2 - x_1}$$
$$= \frac{3 - 2}{-5 - 1}$$
$$= -\frac{1}{6}$$

Slope of BC:

$$\text{slope} = \frac{y_2 - y_1}{x_2 - x_1}$$
$$= \frac{-3 - 3}{-6 - (-5)}$$
$$= 6$$

All four sides are congruent because they have the same lengths. $\overline{AB}$ and $\overline{BC}$ are perpendicular because their slopes are negative reciprocals. Therefore, $ABCD$ is a square.

Topic	Question Numbers	Number of Points	Your Points	Your Percentage
1. Basic Angle and Segment Relationships				
2. Angle and Segment Relationships in Triangles and Polygons	4, 5	4		
3. Constructions	29, 32	6		
4. Transformations	1, 9, 10, 23, 27	10		
5. Triangle Congruence	33	4		
6. Lines, Segments, and Circles on the Coordinate Plane	3, 6, 8, 28	8		
7. Similarity	16, 18	4		
8. Trigonometry	15, 19, 20, 24, 26	10		
9. Parallelograms	2, 7, 13, 25	8		
10. Coordinate Geometry Proofs	35	6		
11. Volume and Solids	11, 14, 21, 31	8		
12. Modeling	17, 34	6		
13. Circles	12, 22, 30	6		

HOW TO CALCULATE YOUR GEOMETRY REGENTS EXAM SCORE

You will need to convert your raw score out of 85 points to a scaled score, which will be your official Geometry Regents grade. Add the total number of credits you scored on the exam to get your raw score. Partial credit is allowed on Parts II, III, and IV so you may need to estimate the number of credits for questions not completely correct. Your scaled score is found using the accompanying conversion chart. Find your raw score in the column labeled "Raw Score". Your final Geometry Regents Exam score is the corresponding number in the "Scaled Score" column. The scaled score is what will be reported to you by your school.

Regents Exam in Geometry—August 2019
Chart for Converting Total Test Raw Scores to
Final Exam Scores (Scale Scores)

Raw Score	Scale Score	Performance Level	Raw Score	Scale Score	Performance Level	Raw Score	Scale Score	Performance Level
80	100	5	53	79	3	26	61	2
79	99	5	52	79	3	25	60	2
78	97	5	51	78	3	24	58	2
77	96	5	50	78	3	23	57	2
76	95	5	49	77	3	22	56	2
75	94	5	48	77	3	21	55	2
74	93	5	47	76	3	20	53	1
73	92	5	46	76	3	19	52	1
72	92	5	45	75	3	18	50	1
71	91	5	44	75	3	17	48	1
70	90	5	43	74	3	16	47	1
69	89	5	42	74	3	15	45	1
68	88	5	41	73	3	14	43	1
67	88	5	40	72	3	13	41	1
66	87	5	39	72	3	12	39	1
65	86	5	38	71	3	11	36	1
64	86	5	37	70	3	10	34	1
63	85	5	36	70	3	9	31	1
62	84	4	35	69	3	8	28	1
61	84	4	34	68	3	7	25	1
60	83	4	33	67	3	6	22	1
59	83	4	32	67	3	5	19	1
58	82	4	31	66	3	4	16	1
57	82	4	30	65	3	3	12	1
56	81	4	29	64	2	2	8	1
55	81	4	28	63	2	1	4	1
54	80	4	27	62	2	0	0	1

January 2020 Exam
Geometry

HIGH SCHOOL MATH REFERENCE SHEET

1 inch = 2.54 centimeters	1 cup = 8 fluid ounces
1 meter = 39.37 inches	1 pint = 2 cups
1 mile = 5280 feet	1 quart = 2 pints
1 mile = 1760 yards	1 gallon = 4 quarts
1 mile = 1.609 kilometers	1 gallon = 3.785 liters
	1 liter = 0.264 gallon
1 kilometer = 0.62 mile	1 liter = 1000 cubic centimeters
1 pound = 16 ounces	
1 pound = 0.454 kilogram	
1 kilogram = 2.2 pounds	
1 ton = 2000 pounds	

Triangle	$A = \frac{1}{2}bh$
Parallelogram	$A = bh$
Circle	$A = \pi r^2$
Circle	$C = \pi d$ or $C = 2\pi r$

General Prism	$V = Bh$
Cylinder	$V = \pi r^2 h$
Sphere	$V = \frac{4}{3}\pi r^3$
Cone	$V = \frac{1}{3}\pi r^2 h$
Pyramid	$V = \frac{1}{3}Bh$
Pythagorean Theorem	$a^2 + b^2 = c^2$
Quadratic Formula	$x = \dfrac{-b \pm \sqrt{b^2 - 4ac}}{2a}$
Arithmetic Sequence	$a_n = a_1 + (n-1)d$
Geometric Sequence	$a_n = a_1 r^{n-1}$
Geometric Series	$S_n = \dfrac{a_1 - a_1 r^n}{1 - r}$ where $r \neq 1$
Radians	$1 \text{ radian} = \dfrac{180}{\pi} \text{ degrees}$
Degrees	$1 \text{ degree} = \dfrac{\pi}{180} \text{ radians}$
Exponential Growth/Decay	$A = A_0 e^{k(t - t_0)} + B_0$

180 Minutes–35 Questions

PART I

Answer all **24** questions in this part. Each correct answer will receive **2** credits. No partial credit will be allowed. For each statement or question, write in the space provided the numeral preceding the word or expression that best completes the statement or answers the question. [48 credits]

1 In the diagram below, $\overleftrightarrow{FAD} \parallel \overleftrightarrow{EHC}$, and $\overline{ABH}$ and $\overline{BC}$ are drawn.

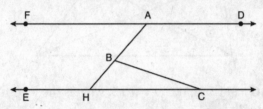

If m∠FAB = 48° and m∠ECB = 18°, what is m∠ABC?

(1) 18°   (3) 66°
(2) 48°   (4) 114°   1 _____

2 A cone has a volume of 108π and a base diameter of 12. What is the height of the cone?

(1) 27   (3) 3
(2) 9   (4) 4   2 _____

3 Triangle *JGR* is similar to triangle *MST*. Which statement is *not* always true?

(1) ∠J ≅ ∠M   (3) ∠R ≅ ∠T
(2) ∠G ≅ ∠T   (4) ∠G ≅ ∠S   3 _____

4 In parallelogram *ABCD*, diagonals $\overline{AC}$ and $\overline{BD}$ intersect at *E*. Which statement proves *ABCD* is a rectangle?

(1) $\overline{AC} \cong \overline{BD}$ (3) $\overline{AC} \perp \overline{BD}$
(2) $\overline{AB} \perp \overline{BD}$ (4) $\overline{AC}$ bisects ∠BCD   4 _____

5 The endpoints of directed line segment PQ have coordinates of $P(-7,-5)$ and $Q(5,3)$. What are the coordinates of point A, on $\overline{PQ}$, that divide PQ into a ratio of $1:3$?

(1) $A(-1,-1)$ (3) $A(3,2)$
(2) $A(2,1)$ (4) $A(-4,-3)$ 5_____

6 In trapezoid $ABCD$ below, $\overline{AD} \parallel \overline{CD}$.

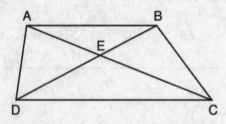

If $AE = 5.2$, $AC = 11.7$, and $CD = 10.5$, what is the length of $\overline{AB}$, to the *nearest tenth*?

(1) 4.7   (3) 8.4
(2) 6.5   (4) 13.1   6_____

7 Kayla was cutting right triangles from wood to use for an art project. Two of the right triangles she cut are shown below.

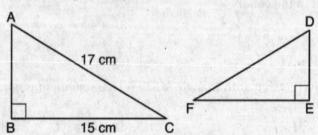

If $\triangle ABC \sim \triangle DEF$, with right angles B and E, $BC = 15$ cm, and $AC = 17$ cm, what is the measure of $\angle F$, to the *nearest degree*?

(1) 28°   (3) 62°
(2) 41°   (4) 88°   7_____

8 The line represented by $2y = x + 8$ is dilated by a scale factor of k centered at the origin, such that the image of the line has an equation of $y - \frac{1}{2}x = 2$. What is the scale factor?

(1) $k = \frac{1}{2}$ (3) $k = \frac{1}{4}$

(2) $k = 2$ (4) $k = 4$ 8 _____

9 In quadrilateral $ABCD$ below, $\overline{AB} \parallel \overline{CD}$, and E, H, and F are the midpoints of $\overline{AD}$, $\overline{AC}$, and $\overline{BC}$, respectively.

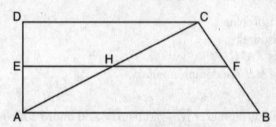

If $AB = 24$, $CD = 18$, and $AH = 10$, then FH is

(1) 9   (3) 12

(2) 10   (4) 21   9 _____

10 Jaden is comparing two cones. The radius of the base of cone A is twice as large as the radius of the base of cone B. The height of cone B is twice the height of cone A. The volume of cone A is

(1) twice the volume of cone B

(2) four times the volume of cone B

(3) equal to the volume of cone B

(4) equal to half the volume of cone B 10 _____

11 A regular hexagon is rotated about its center. Which degree measure will carry the regular hexagon onto itself?

(1) $45°$ (3) $120°$

(2) $90°$ (4) $135°$ 11 _____

12 In triangle *MAH* below, $\overline{MT}$ is the perpendicular bisector of $\overline{AH}$.

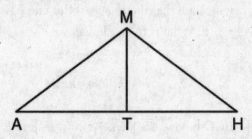

Which statement is *not* always true?

(1) $\triangle MAH$ is isosceles.
(2) $\triangle MAT$ is isosceles.
(3) $\overline{MT}$ bisects $\angle AMH$.
(4) $\angle A$ and $\angle TMH$ are complementary.

12 _____

13 In circle *B* below, diameter $\overline{RT}$, radius $\overline{BE}$, and chord $\overline{RE}$ are drawn.

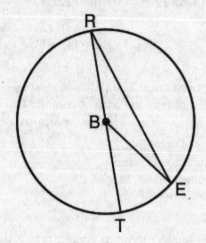

If m$\angle TRE = 15°$ and $BE = 9$, then the area of sector *EBR* is

(1) 3.375π (3) 33.75π
(2) 6.75π (4) 37.125π

13 _____

14 Lou has a solid clay brick in the shape of a rectangular prism with a length of 8 inches, a width of 3.5 inches, and a height of 2.25 inches. If the clay weighs 1.055 oz/in^3, how much does Lou's brick weigh, to the *nearest ounce*?

(1) 66   (3) 63

(2) 64   (4) 60   14 _____

15 Rhombus *ABCD* can be mapped onto rhombus *KLMN* by a rotation about point *P*, as shown below.

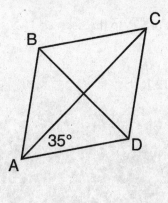 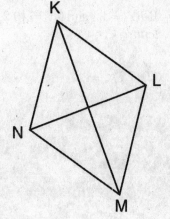

What is the measure of ∠*KNM* if the measure of ∠*CAD* = 35°?

(1) 35°   (3) 70°

(2) 55°   (4) 110°   15 _____

16 In right triangle *RST* below, altitude $\overline{SV}$ is drawn to hypotenuse $\overline{RT}$.

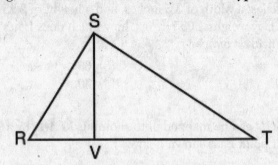

If $RV = 4.1$ and $TV = 10.2$, what is the length of $\overline{ST}$, to the *nearest tenth*?

(1) 6.5   (3) 11.0

(2) 7.7   (4) 12.1   16 _____

17 On the set of axes below, pentagon *ABCDE* is congruent to
   A″B″C″D″E″.

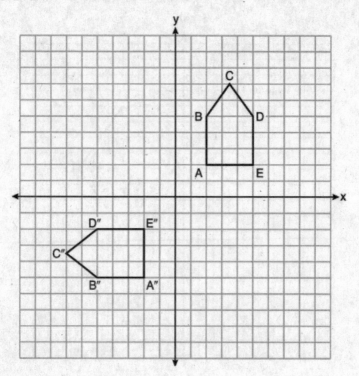

Which describes a sequence of rigid motions that maps *ABCDE*
onto *A″B″C″D″E″*?

(1) a rotation of 90° counterclockwise about the origin followed by
    a reflection over the *x*-axis

(2) a rotation of 90° counterclockwise about the origin followed by
    a translation down 7 units

(3) a reflection over the *y*-axis followed by a reflection over the *x*-axis

(4) a reflection over the *x*-axis followed by a rotation of 90°
    counterclockwise about the origin   17 _____

18 On the set of axes below, rhombus $ABCD$ has vertices whose coordinates are $A(1,2)$, $B(4,6)$, $C(7,2)$, and $D(4,-2)$.

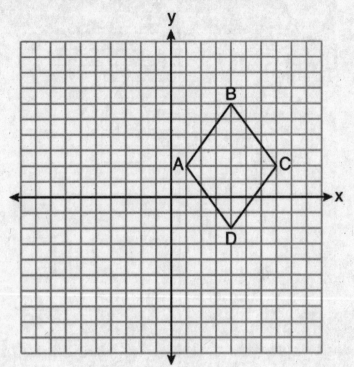

What is the area of rhombus $ABCD$?

(1) 20   (3) 25

(2) 24   (4) 48   18 ____

19 Which figure(s) below can have a triangle as a two-dimensional cross section?

I. cone
II. cylinder
III. cube
IV. square pyramid

(1) I, only   (3) I, II, and IV, only

(2) IV, only   (4) I, III, and IV, only   19 ____

20 What is an equation of a circle whose center is at $(2,-4)$ and is tangent to the line $x = -2$?

(1) $(x - 2)^2 + (y + 4)^2 = 4$
(2) $(x - 2)^2 + (y + 4)^2 = 16$
(3) $(x + 2)^2 + (y - 4)^2 = 4$
(4) $(x + 2)^2 + (y - 4)^2 = 16$

20 _____

21 For the acute angles in a right triangle, $\sin(4x)° = \cos(3x + 13)°$. What is the number of degrees in the measure of the *smaller* angle?

(1) $11°$　　　　　　　　(3) $44°$
(2) $13°$　　　　　　　　(4) $52°$

21 _____

22 Triangle PQR is shown on the set of axes below.

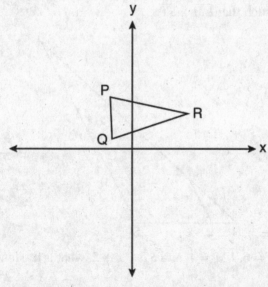

Which quadrant will contain point R'', the image of point R, after a $90°$ clockwise rotation centered at $(0,0)$ followed by a reflection over the x-axis?

(1) I　　　　　　　　(3) III
(2) II　　　　　　　　(4) IV

22 _____

23 In the diagram below of right triangle ABC, altitude $\overline{BD}$ is drawn.

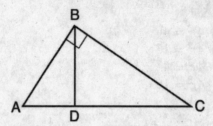

What ratio is always equivalent to cos A?

(1) $\dfrac{AB}{BC}$

(3) $\dfrac{BD}{AB}$

(2) $\dfrac{BD}{BC}$

(4) $\dfrac{BC}{AC}$

23 _____

24 In the diagram below of $\triangle RST$, L is a point on $\overline{RS}$, and M is a point on $\overline{RT}$, such that $\overline{LM} \parallel \overline{ST}$.

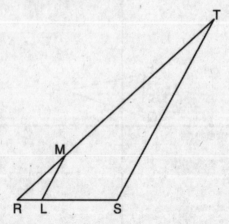

If $RL = 2$, $LS = 6$, $LM = 4$, and $ST = x + 2$, what is the length of $\overline{ST}$?

(1) 10

(3) 14

(2) 12

(4) 16

24 _____

PART II

Answer all 7 questions in this part. Each correct answer will receive 2 credits. Clearly indicate the necessary steps, including appropriate formula substitutions, diagrams, graphs, charts, etc. For all questions in this part, a correct numerical answer with no work shown will receive only 1 credit. [14 credits]

25 In the diagram below, right triangle PQR is transformed by a sequence of rigid motions that maps it onto right triangle NML.

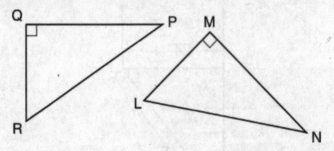

Write a set of three congruency statements that would show ASA congruency for these triangles.

26 Diego needs to install a support beam to hold up his new bird-house, as modeled below. The base of the birdhouse is $24\frac{1}{2}$ inches long. The support beam will form an angle of 38° with the vertical post. Determine and state the approximate length of the support beam, x, to the *nearest inch*.

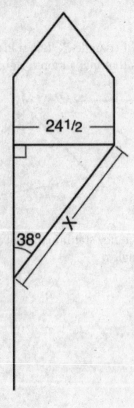

27 A rectangular tabletop will be made of maple wood that weighs 43 pounds per cubic foot. The tabletop will have a length of eight feet, a width of three feet, and a thickness of one inch. Determine and state the weight of the tabletop, in pounds.

28 In the diagram below of circle O, secant $\overline{ABC}$ and tangent $\overrightarrow{AD}$ are drawn.

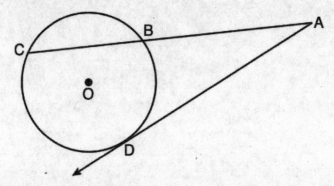

If $CA = 12.5$ and $CB = 4.5$, determine and state the length of $\overline{DA}$.

29 Given $\overline{MT}$ below, use a compass and straightedge to construct a 45° angle whose vertex is at point M.

[Leave all construction marks.]

M ─────────────────────────────────────── T

30 In $\triangle XYZ$ shown below, medians $\overline{XE}$, $\overline{YF}$, and $\overline{ZD}$ intersect at C.

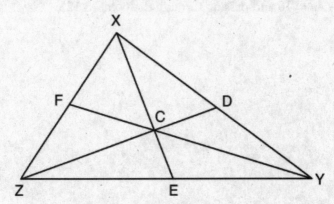

If $CE = 5$, $YF = 21$, and $XZ = 15$, determine and state the perimeter of triangle CFX.

31 Determine and state an equation of the line perpendicular to the line $5x - 4y = 10$ and passing through the point $(5,12)$.

PART III

Answer all 3 questions in this part. Each correct answer will receive 4 credits. Clearly indicate the necessary steps, including appropriate formula substitutions, diagrams, graphs, charts, etc. Utilize the information provided for each question to determine your answer. Note that diagrams are not necessarily drawn to scale. For all questions in this part, a correct numerical answer with no work shown will receive only 1 credit. All answers should be written in pen, except for graphs and drawings, which should be done in pencil. [12 credits]

32 Quadrilateral *NATS* has coordinates $N(-4,-3)$, $A(1,2)$, $T(8,1)$, and $S(3,-4)$.

Prove quadrilateral *NATS* is a rhombus.

[The use of the set of axes below is optional.]

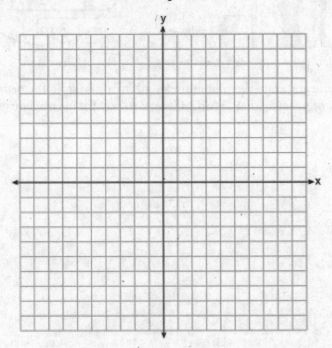

33 David has just finished building his treehouse and still needs to buy
a ladder to be attached to the ledge of the treehouse and anchored
at a point on the ground, as modeled below. David is standing
1.3 meters from the stilt supporting the treehouse. This is the point
on the ground where he has decided to anchor the ladder. The
angle of elevation from his eye level to the bottom of the treehouse
is 56 degrees. David's eye level is 1.5 meters above the ground.

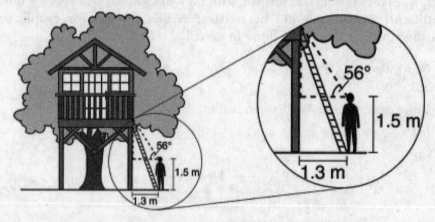

Determine and state the minimum length of a ladder, to the *near-est tenth of a meter*, that David will need to buy for his treehouse.

34 A manufacturer is designing a new container for their chocolate-covered almonds. Their original container was a cylinder with a height of 18 cm and a diameter of 14 cm. The new container can be modeled by a rectangular prism with a square base and will contain the same amount of chocolate-covered almonds.

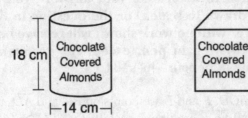

If the new container's height is 16 cm, determine and state, to the *nearest tenth of a centimeter*, the side length of the new container if both containers contain the same amount of almonds.

A store owner who sells the chocolate-covered almonds displays them on a shelf whose dimensions are 80 cm long and 60 cm wide. The shelf can only hold one layer of new containers when each new container sits on its square base. Determine and state the maximum number of new containers the store owner can fit on the shelf.

PART IV

Answer the question in this part. A correct answer will receive 6 credits. Clearly indicate the necessary steps, including appropriate formula substitutions, diagrams, graphs, charts, etc. Utilize the information provided for the question to determine your answer. Note that diagrams are not necessarily drawn to scale. For the question in this part, a correct numerical answer with no work shown will receive only 1 credit. All answers should be written in pen, except for graphs and drawings, which should be done in pencil.   [6 credits]

35 In quadrilateral $ABCD$, E and F are points on $\overline{BC}$ and $\overline{AD}$, respectively, and $\overline{BGD}$ and $\overline{EGF}$ are drawn such that $\angle ABG \cong \angle CDG$, $\overline{AB} \cong \overline{CD}$, and $\overline{CE} \cong \overline{AF}$.

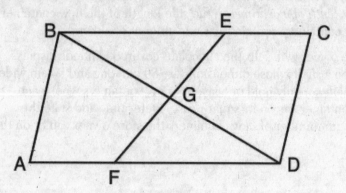

Prove: $\overline{FG} \cong \overline{EG}$

Answers January 2020
Geometry

Answer Key

PART I

1. 3	5. 4	9. 3	13. 3	17. 2	21. 3
2. 2	6. 3	10. 1	14. 1	18. 2	22. 1
3. 2	7. 1	11. 3	15. 4	19. 4	23. 2
4. 1	8. 1	12. 2	16. 4	20. 2	24. 4

PART II

25. See the detailed solution for the set of three congruency statements.

26. 40. Sine calculates the ratio of the length of the opposite side of a right triangle to the length of the hypotenuse.

27. $2 \text{ ft}^3 \times 43 \text{ lb/ft}^3 = 86 \text{ lb}$

28. $12.5 \times 8 = DA^2$, so $DA^2 = 100$ and $DA = 10$

29. See the detailed solution for the construction.

30. $10 + 7 + 7.5 = 24.5$

31. $y - 12 = -\frac{4}{5}(x - 5)$

PART III

32. *NATS* is a rhombus because all of the side lengths of the quadrilateral are the same.

33. 3.7 meters

34. 13.2 cm, 24 containers

PART IV

35. See the detailed solution for the proof.

In **PARTS II–IV**, you are required to show how you arrived at your answers. For sample methods of solutions, see the *Answers Explained* section.

Answers Explained

PART I

1. Because $\overline{FAD}$ and $\overline{EHC}$ are parallel lines, $\overline{FAB}$ and $\overline{BHC}$ are congruent. Consequently, m∠BHC = 48°. Because the interior angles of a triangle sum to 180°, m∠BHC + m∠HCB + m∠CBH = 180°. Additionally, ∠ECB is congruent to ∠HCB because they are the same angle. Consequently, 18° + 48° + x = 180°, where x represents the angle measure for ∠CBH. Therefore, m∠CBH = 114°. Because ∠CBH and ∠ABC are supplementary, m∠CBH + m∠ABC = 180°.

Therefore, m∠ABC = 180° − 114° = 66°.

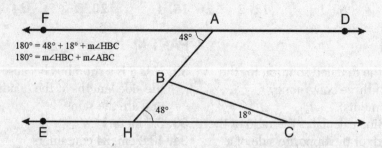

The correct choice is **(3)**.

2. For a right circular cone, the volume is given by $V = \frac{1}{3}\pi r^2 h$. The diameter of 12 means that the radius is 6 because the diameter is equal to 2 times the radius. Consequently, $108\pi = \frac{1}{3}\pi 6^2 h$. Cancel out π from both sides and expand 6^2. Hence, $108 = \frac{1}{3} \cdot 36 \cdot h$, which gives $108 = 12h$. So $h = 9$.

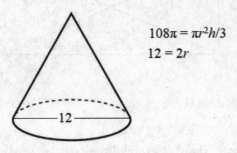

$108\pi = \pi r^2 h/3$

$12 = 2r$

The correct choice is (**2**).

3. When triangles are similar, their corresponding angle measures are congruent. Since triangle JGR is similar to triangle MST, $\angle J \cong \angle M$, $\angle G \cong \angle S$, and $\angle R \cong \angle T$. The only remaining statement that does not correspond to one of these statements is $\angle G \cong \angle T$, which is not necessarily true.

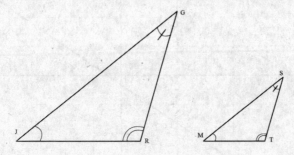

The correct choice is (**2**).

4. If the two diagonals of a parallelogram are congruent, then the parallelogram is a rectangle.

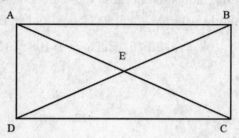

The correct choice is (1).

5. In the x-coordinate, there is a total distance between P and Q of 12 units. In the y-coordinate, there is a total distance between P and Q of 8 units. Fortunately, these divide nicely into 3 : 9 and 2 : 6 to maintain the desired ratio of 1 : 3. To find point A, add 3 units to the x-coordinate of P and 2 units to the y-coordinate of P. Hence, $(-7 + 3, -5 + 2) = (-4, -3)$.

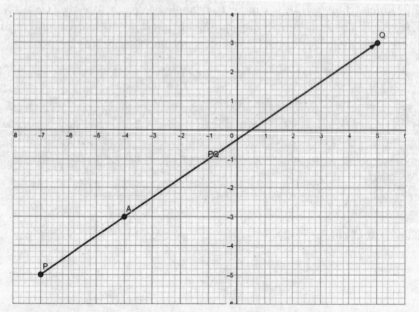

The correct choice is (4).

6. In a trapezoid like $ABCD$ with the diagonals, the resulting triangles AEB and CED are similar. Since AC is 11.7 and AE is 5.2, EC is $11.7 - 5.2 = 6.5$. So triangle AEB is in a $6.5 : 5.2$ ratio with triangle CED. This means that $CD : AB = 6.5 : 5.2$ and hence $10.5 : AB = 6.5 : 5.2$. This can be rewritten as $\dfrac{10.5}{AB} = \dfrac{6.5}{5.2}$, which means that $AB = 8.4$.

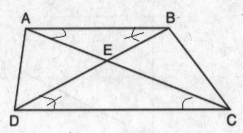

The correct choice is **(3)**.

7. Since triangles ABC and DEF are similar, angles C and F are congruent. For angle C, $\cos C = \dfrac{15}{17}$. Hence, $\arccos\left(\dfrac{15}{17}\right) = C \approx 28°$.

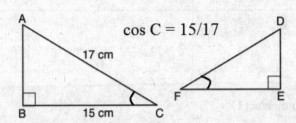

The correct choice is **(1)**.

8. When dilating a line, the slope remains the same and the only part that is impacted is the y-intercept when the lines are written in $y = mx + b$ format. The original line can be written as $y = \frac{x}{2} + 4$ and the dilated line can be written as $y = \frac{x}{2} + 2$. Hence, the dilation factor can be obtained by comparing the dilated y-intercept with the original y-intercept to give $k = \frac{2}{4} = \frac{1}{2}$.

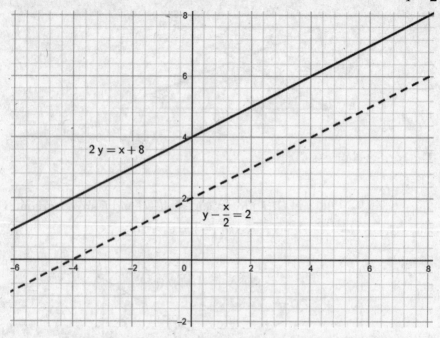

The correct choice is (**1**).

9. The triangles ACB and HCF are similar because they share a common angle and the ratio between AC and HC and the ratio between BC and FC is $2 : 1$. Therefore, the ratio between AB and FH is also $2 : 1$. Since $AB = 24$, $FH = \dfrac{24}{2} = 12$.

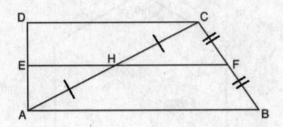

The correct choice is (**3**).

10. For a right circular cone, the volume is given by $\frac{1}{3}\pi r^2 h$. For cone A, the volume is $\frac{1}{3}\pi r_A{}^2 h_A$, where $r_A = 2r_B$ is the radius of cone A and $h_A = \frac{1}{2}h_B$ is the height of cone A. For cone B, the volume is $\frac{1}{3}\pi r_B{}^2 h_B$. By making the substitutions $r_A = 2r_B$ and $h_A = \frac{1}{2}h_B$ in the formula for the volume of cone A, $\frac{1}{3}\pi r_A{}^2 h_A = \frac{1}{3}\pi (2r_B)^2\left(\frac{1}{2}h_B\right)$. This simplifies to $\frac{1}{3}\pi (4r_B)\left(\frac{1}{2}h_B\right)$, which becomes $\frac{1}{3}\cdot 2\pi r_B h_B$. So the volume of cone A is twice the volume of cone B.

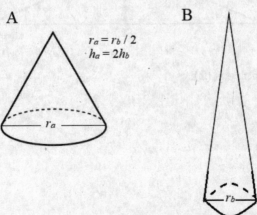

The correct choice is (**1**).

11. The rotation will happen about the center of the regular hexagon. Note that the interior can be divided into six 60° sections from the center of the regular hexagon, so any multiple of 60° will rotate the regular hexagon back onto itself. Hence, 120° is the answer, since 120° = 2 × 60°.

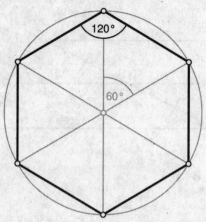

The correct choice is (**3**).

12. Triangle *MAT* may not be isosceles because *MT* could be much smaller (or larger) than *TA* while still being the perpendicular bisector of $\overline{AH}$.

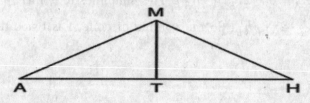

The correct choice is (**2**).

13. The formula for the area of a sector is $A = \frac{1}{2}\theta r^2$, where θ is the angle measure in radians. Since triangle RBE is an isosceles triangle, $\angle BRE$ and $\angle BER$ are congruent and the measure of each is equal to 15°. Hence, $m\angle RBE = 180° - 15° - 15° = 150° = \frac{5\pi}{6}$. Therefore, $A = 12\frac{15\pi}{6} 9^2 = 33.75\pi$.

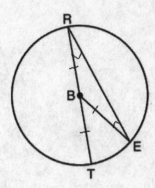

The correct choice is **(3)**.

14. The volume of the brick is 8 in $\times$ 3.5 in $\times$ 2.25 in = 63 in^3. The weight is given by multiplying the density by the volume: 63 in^3 $\times$ 1.055 oz/in^3 = 66.465 oz.

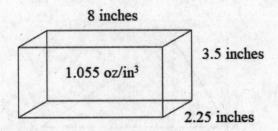

8 inches

3.5 inches

1.055 oz/in^3

2.25 inches

The correct choice is **(1)**.

15. Since a rhombus has four equal sides, triangle KNM is isosceles. $\angle CAD$ corresponds to $\angle NKM$ through the rotation, so m$\angle NKM = 35°$. Since triangle KNM is isosceles, m$\angle KMN = 35°$. Hence, m$\angle KNM = 180° - 35° - 35° = 110°$.

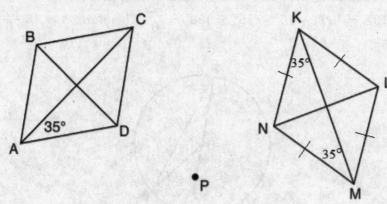

The correct choice is (4).

16. Triangles RST, RVS, and SVT are all similar. Therefore, $\dfrac{RV}{SV} = \dfrac{SV}{TV}$. Substituting the given values for RV and TV gives $\dfrac{4.1}{SV} = \dfrac{SV}{10.2}$, which simplifies to $SV^2 = 41.82$ and $SV \approx 6.5$. Using the Pythagorean theorem for triangle SVT shows that $ST^2 = 6.5^2 + 10.2^2 = 146.29$. So $ST \approx 12.1$.

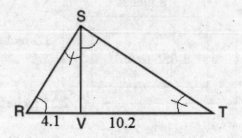

The correct choice is (4).

7. Although choice (1) brings the shape *ABCDE* directly onto the shape *A″B″C″D″E″*, it does not preserve the orientation in which *A* maps to *A″*, *B* maps to *B″*, and so on. Consider the movements that are required to map *A* to *A″*. Note that choice (3) would map *A* to *E″* (and put the shape in the wrong orientation). Choice (4) similarly maps *A* to *E″*. To check choice (2), perform the rotation to obtain *A′B′C′D′E′* and note that a translation down 7 units would map *A′B′C′D′E′* directly onto the shape *A″B″C″D″E″* with the proper orientation.

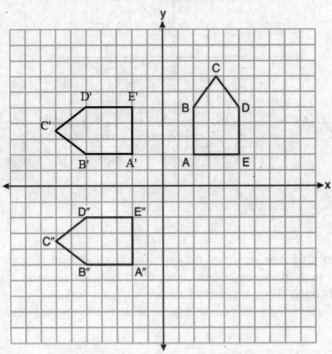

The correct choice is **(2)**.

18. The area of a rhombus is $\frac{pq}{2}$, where p and q are the lengths of the diagonals In this case, the area of $ABCD$ is $AC \times \frac{BD}{2} = 6 \times \frac{8}{2} = 24$.

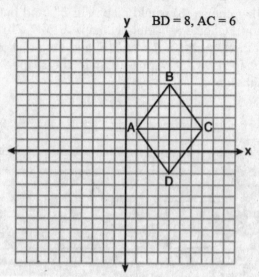

BD = 8, AC = 6

The correct choice is **(2)**.

19. To obtain a triangle as a two-dimensional cross section of a cone, take any plane through the vertex of the cone and its base. For a cube, take any plane angled through three surfaces of the cube. For a square pyramid, take a triangular face of the pyramid. It is impossible to obtain such a triangle with a cylinder.

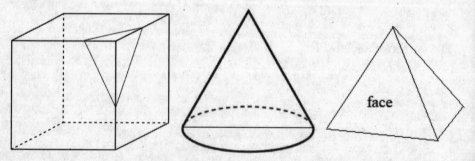

face

The correct choice is **(4)**.

20. The distance between the center point at $(2, -4)$ and the line $x = -2$ must be 4. Therefore, the radius of the circle is 4. Recall that the standard form for a circle is $(x - h)^2 + (y - k)^2 = r^2$, where (h, k) is the center and r is the radius. Therefore, the standard form for this circle is $(x - 2)^2 + (y + 4)^2 = 16$.

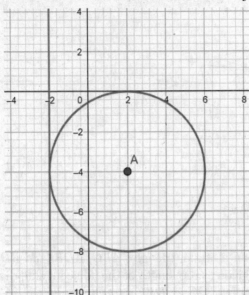

The correct choice is (2).

21. Since the two angles are acute angles in a right triangle, $4x + (3x + 13) = 90$. Therefore, $7x = 77$, which means that $x = 11$. Hence, the two angles are $4x = 44°$ and $3x + 13 = 46°$. Hence, the smaller angle is $44°$.

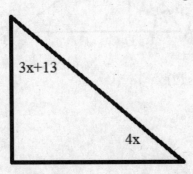

The correct choice is (3).

22. The 90° clockwise rotation maps R to R' in quadrant IV. Then the reflection over the x-axis maps R' to R'' in quadrant I.

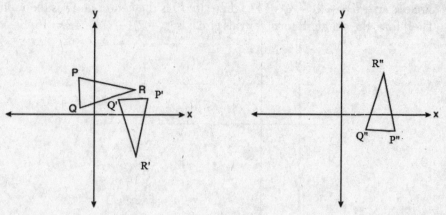

The correct choice is **(1)**.

23. Because $\overline{BD}$ is an altitude, triangles ABC, ADB, and BDC are similar. The cosine of $\angle A$ is given as the length of the adjacent side divided by the hypotenuse. In triangle BDC, $\angle DBC$ is congruent to $\angle A$ (that is, $\angle BAC$). Therefore, the adjacent side ($\overline{BD}$) divided by the hypotenuse ($\overline{BC}$) would give $\cos DBC = \cos A$. Hence, $\cos A = \dfrac{BD}{BC}$.

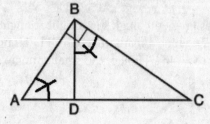

The correct choice is **(2)**.

24. Because they share an angle and parallel bases, triangles RST and RLM are similar. Therefore, $\frac{RL}{RS} = \frac{LM}{ST}$. Note that $RS = RL + RS$, so $\frac{RL}{(RL + RS)}$ $= \frac{LM}{ST}$. With the relevant values inserted, $\frac{2}{8} = \frac{4}{(x+2)}$. Hence, $x = 14$, and $ST = 14 + 2 = 16$.

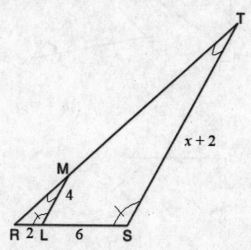

The correct choice is **(4)**.

PART II

25. $Q \cong \angle M, \angle P \cong \angle N, \overline{QP} \cong \overline{MN}$

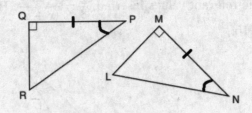

$\angle Q \cong \angle M, \angle R \cong \angle L, \overline{QR} \cong \overline{ML}$

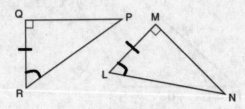

$\angle P \cong \angle N, \angle R \cong \angle L, \overline{PR} \cong \overline{NL}$

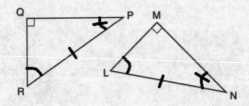

26. Recall that sine calculates the ratio of the length of the opposite side of a right triangle to the length of the hypotenuse. Therefore, $\sin 38° = \dfrac{24.5}{x}$. Solving for x gives $x = \dfrac{24.5}{\sin 38°} = \dfrac{24.5}{0.61566147532} = 39.7945965147 \approx$ 40 inches.

27. The volume of the tabletop is $8 \text{ ft} \times 3 \text{ ft} \times \frac{1}{12} \text{ ft} = 2 \text{ ft}^3$. The weight is given by multiplying the density by the volume: $2 \text{ ft}^3 \times 43 \text{ lb/ft}^3 = 86 \text{ lb}$.

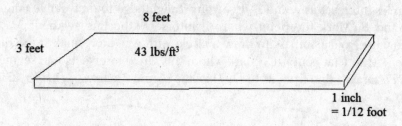

28. When a secant and a tangent of a circle are drawn from a common point outside a circle, the product of the lengths of the secant and its external segment length is equal to the square of the tangent segment length. In this case, $CA \times BA = DA^2$. Note that $BA = CA - CB = 12.5 - 4.5 = 8$. Hence, $12.5 \times 8 = DA^2$. So $DA^2 = 100$ and $DA = 10$.

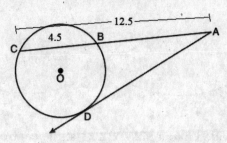

29. To begin, bisect segment $\overline{MT}$. From point M, use your compass to draw a circle with a radius that is at least half the distance between M and T. Then draw a circle with the same radius at center T. Mark the points where these circles intersect as A and B. Use your straightedge to draw a line connecting A and B. Mark the point where it intersects the line segment $\overline{MT}$ as C. Now use your compass to draw a circle with a radius equal to the distance from M to C at center C. Mark where this circle intersects line segment $\overline{AB}$ as D. Draw a line from M to D. This line segment makes $\angle TMD$ a 45° angle.

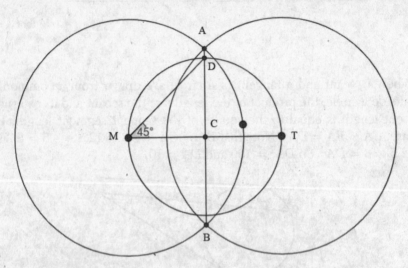

30. Medians are divided into a ratio of 2 : 1 at the centroid. That is, the part of the median nearer to the vertex is twice the length of the part nearer to the side. In this case, that means that $CY = 2CF$, $XC = 2CE$, and $ZC = 2CD$. Because $CE = 5$, $XC = 10$. Because $YF = 21$, $CY = 14$ and $FC = 7$. Because F is the median of $\overline{XZ}$ and $XZ = 15$, $XF = 7.5$. The perimeter of triangle $CFX = 10 + 7 + 7.5 = 24.5$.

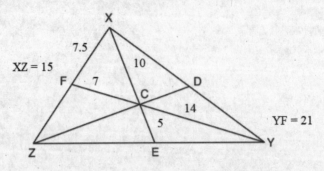

31. First write the equation $5x - 4y = 10$ in slope-intercept format to obtain $y = \frac{5}{4}x - \frac{5}{2}$. Therefore, the original line has a slope of $\frac{5}{4}$, so the line perpendicular to this line will have a slope of $-\frac{4}{5}$. To find the line with this slope passing through $(5, 12)$, use the point-slope form of a line to write $y - 12 = -\frac{4}{5}(x - 5)$.

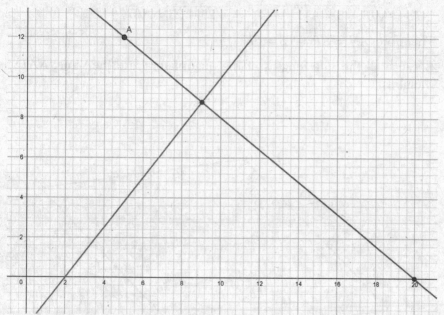

PART III

32. To show that *NATS* is a rhombus, show that each side is the same length.

$$AN = \sqrt{(1 - -4)^2 + (2 - -3)^2} = \sqrt{25 + 25} = \sqrt{50}$$

$$AT = \sqrt{(8 - 1)^2 + (1 - 2)^2} = \sqrt{49 + 1} = \sqrt{50}$$

$$SN = \sqrt{(3 - -4)^2 + (-4 - -3)^2} = \sqrt{49 + 1} = \sqrt{50}$$

$$TS = \sqrt{(8 - 3)^2 + (1 - -4)^2} = \sqrt{25 + 25} = \sqrt{50}$$

Therefore, $AN = AT = TS = SN$.

Because all the side lengths of the quadrilateral are the same, *NATS* is a rhombus.

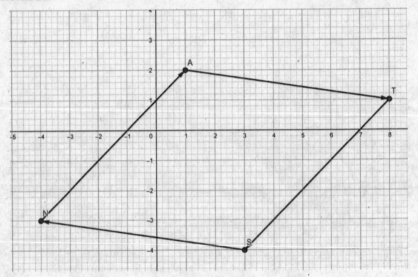

33. To determine the length of the ladder, first find the distance from the ground to the treehouse. David's eye is already 1.5 meters above the ground. The distance from David's eye level to the treehouse is marked x on the graph. Note that $\tan 56° = \dfrac{x}{1.3}$, which means that $x = 1.3 \tan 56° = 1.927329$ meters.

So the entire height is 1.5 meters + 1.927329 meters = 3.427329 meters. Then use the Pythagorean theorem to obtain the ladder length: $c^2 = 3.4273292^2 + 1.32^2$ and $c \approx 3.7$ meters.

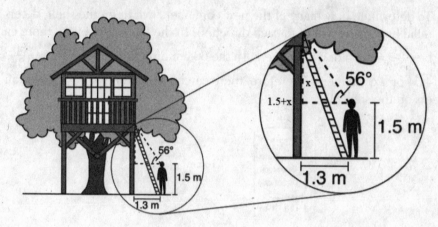

34. The volume for a right cylinder is $V = \pi r^2 h$. The radius of the cylinder is half the diameter, so $r = \dfrac{d}{2} = \dfrac{14}{2} = 7$ cm. Consequently, the volume of the cylinder is $\pi(7^2)(18) = 882\pi$ cm^3. The volume of the rectangular prism is $V = 16s^2$, where s is the side length of the square base. Set the volumes equal to each other and solve for s to obtain $882\pi = 16s^2$, which means that $s^2 = \dfrac{882\pi}{16} = 173.18$. Take the square root of both sides to obtain $s = 13.2$ cm.

To determine how many of the new containers will fit on the shelf, determine how many will fit in each direction. In the 80 cm direction, there can be $\dfrac{80}{13.2} = 6.06$ or 6 containers. In the 60 cm direction, there can be $\dfrac{60}{13.2} = 4.54$ or 4 containers. Therefore, there can be a total of $6 \times 4 = 24$ containers on the shelf.

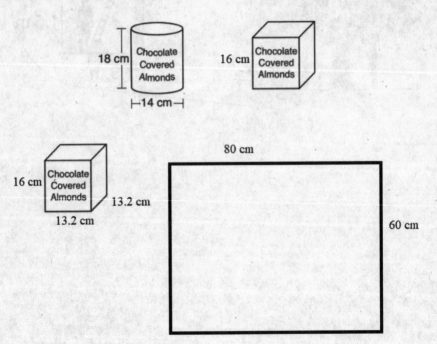

PART IV

35. Given $\angle ABG \cong \angle CDG$ and $\overline{AB} \cong \overline{CD}$ and since $\overline{BD} \cong \overline{BD}$, triangles ABD and CDB are congruent by side-angle-side. Therefore, $\overline{BC}$ is congruent to $\overline{DA}$ because corresponding parts of congruent triangles are congruent.

Since $\overline{BC} \cong \overline{DA}$ and given $\overline{CE} \cong \overline{AF}$, $\overline{BE}$ and $\overline{DF}$ must be congruent because $BE = BC - CE = DA - AF = DF$. Note that $\angle BGE \cong \angle DGF$ because they are vertical angles.

Additionally, $\angle DBC \cong \angle BDA$ because corresponding parts of the congruent triangles ABD and CDB are congruent. Therefore, triangles GBE and GDF are congruent by angle-angle-side. As a result, $\overline{FG} \cong \overline{EG}$ because corresponding parts of congruent triangles are congruent.

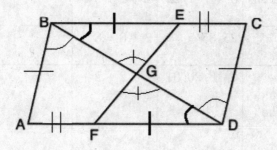

Topic	Question Numbers	Number of Points	Your Points	Your Percentage
1. Basic Angle and Segment Relationships	1	2		
2. Angle and Segment Relationships in Triangles and Polygons	12, 30	$2 + 2 = 4$		
3. Constructions	29	2		
4. Transformations	8, 11, 17, 22	$2 + 2 + 2 + 2 = 8$		
5. Triangle Congruence	25, 35	$2 + 6 = 8$		
6. Lines, Segments, and Circles on the Coordinate Plane	5, 18, 20, 31	$2 + 2 + 2 + 2 = 8$		
7. Similarity	3, 6, 16, 24	$2 + 2 + 2 + 2 = 8$		
8. Trigonometry	7, 21, 23, 26, 33	$2 + 2 + 2 + 2 + 4 = 12$		
9. Parallelograms	4, 9, 15	$2 + 2 + 2 = 6$		
10. Coordinate Geometry Proofs	32	4		
11. Volume and Solids	2, 10, 19, 34	$2 + 2 + 2 + 4 = 10$		
12. Modeling	14, 27	$2 + 2 = 4$		
13. Circles	13, 28	$2 + 2 = 4$		

HOW TO CALCULATE YOUR GEOMETRY REGENTS EXAM SCORE

You will need to convert your raw score out of 85 points to a scaled score, which will be your official Geometry Regents grade. Add the total number of credits you scored on the exam to get your raw score. Partial credit is allowed on Parts II, III, and IV so you may need to estimate the number of credits for questions not completely correct. Your scaled score is found using the accompanying conversion chart. Find your raw score in the column labeled "Raw Score". Your final Geometry Regents Exam score is the corresponding number in the "Scaled Score" column. The scaled score is what will be reported to you by your school.

Regents Exam in Geometry—January 2020
Chart for Converting Total Test Raw Scores to
Final Exam Scores (Scale Scores)

Raw Score	Scale Score	Performance Level	Raw Score	Scale Score	Performance Level	Raw Score	Scale Score	Performance Level
80	100	5	53	79	3	26	60	2
79	99	5	52	79	3	25	59	2
78	98	5	51	78	3	24	57	2
77	97	5	50	78	3	23	56	2
76	95	5	49	77	3	22	55	2
75	94	5	48	77	3	21	53	1
74	93	5	47	76	3	20	52	1
73	92	5	46	76	3	19	50	1
72	92	5	45	75	3	18	48	1
71	91	5	44	75	3	17	46	1
70	90	5	43	74	3	16	44	1
69	89	5	42	73	3	15	42	1
68	88	5	41	73	3	14	40	1
67	87	5	40	72	3	13	38	1
66	87	5	39	72	3	12	36	1
65	86	5	38	71	3	11	34	1
64	86	5	37	70	3	10	31	1
63	85	5	36	70	3	9	29	1
62	84	4	35	69	3	8	26	1
61	83	4	34	68	3	7	23	1
60	83	4	33	67	3	6	20	1
59	82	4	32	66	3	5	17	1
58	82	4	31	65	3	4	14	1
57	81	4	30	64	2	3	11	1
56	81	4	29	63	2	2	7	1
55	80	4	28	62	2	1	4	1
54	80	4	27	61	2	0	0	1

Sample 2025 Exam
Geometry

HIGH SCHOOL MATH REFERENCE SHEET

1 inch = 2.54 centimeters

1 meter = 39.37 inches

1 mile = 5280 feet

1 mile = 1760 yards

1 mile = 1.609 kilometers

1 kilometer = 0.62 mile

1 pound = 16 ounces

1 pound = 0.454 kilogram

1 kilogram = 2.2 pounds

1 ton = 2000 pounds

1 cup = 8 fluid ounces

1 pint = 2 cups

1 quart = 2 pints

1 gallon = 4 quarts

1 gallon = 3.785 liters

1 liter = 0.264 gallon

1 liter = 1000 cubic centimeters

Triangle	$A = \dfrac{1}{2}bh$
Parallelogram	$A = bh$
Circle	$A = \pi r^2$
Circle	$C = \pi d$ or $C = 2\pi r$

General Prism	$V = Bh$
Cylinder	$V = \pi r^2 h$
Sphere	$V = \frac{4}{3}\pi r^3$
Cone	$V = \frac{1}{3}\pi r^2 h$
Pyramid	$V = \frac{1}{3}Bh$
Pythagorean Theorem	$a^2 + b^2 = c^2$
Quadratic Formula	$x = \dfrac{-b \pm \sqrt{b^2 - 4ac}}{2a}$
Arithmetic Sequence	$a_n = a_1 + (n-1)d$
Geometric Sequence	$a_n = a_1 r^{n-1}$
Geometric Series	$S_n = \dfrac{a_1 - a_1 r^n}{1 - r}$ where $r \neq 1$
Radians	$1 \text{ radian} = \dfrac{180}{\pi}$ degrees
Degrees	$1 \text{ degree} = \dfrac{\pi}{180}$ radians
Exponential Growth/Decay	$A = A_0 e^{k(t - t_0)} + B_0$

180 Minutes–35 Questions

PART I

Answer all 24 questions in this part. Each correct answer will receive 2 credits. Utilize the information provided for each question to determine your answer. Note that diagrams are not necessarily drawn to scale. No partial credit will be allowed.   [48 credits]

1 A cylinder has a height of 30 inches and circular bases with a diameter of 25 inches. If a plane intersects the cylinder perpendicular to the circular bases and passes through diameter $\overline{AB}$, what cross section is intercepted?

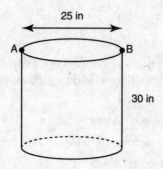

(1) an isosceles triangle with legs measuring 30 inches and a base of 25 inches

(2) a square with a side length of 25 inches

(3) a rectangle with a width of 25 inches and a length of 30 inches

(4) a circle with a diameter of 25 inches   1 _____

2 Juaquin wants to prove line p is parallel to line m in the figure
below. He can do this by showing $\angle 1$ is congruent to which other
angle?

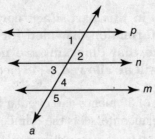

(1) 2   (3) 4

(2) 3   (4) 5   2 _____

3 A quadrilateral has two pairs of parallel opposite sides. Which
piece of additional information would be sufficient to prove the
quadrilateral is a rectangle?

(1) opposite sides are congruent

(2) the diagonals are perpendicular

(3) the diagonals are congruent

(4) opposite angles are congruent   3 _____

4 What is the image of the point (5,1) after a reflection through the
point (3,4)?

(1) (7,−2)   (3) (4,2.5)

(2) (1,7)   (4) (8,5)   4 _____

5 The pyramid shown below has a rectangular base with a width of 26 centimeters and a length of 27 centimeters. The height of the pyramid is 14 centimeters.

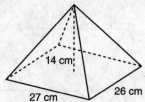

What is the volume of the pyramid?

(1) 3276 cm^3

(3) 4914 cm^3

(2) 3403 cm^3

(4) 9828 cm^3

5 _____

6 In the figure below, side $\overline{BG}$ of $\triangle BIG$ is extended to L. If $m\angle B = (x + 28)°$, $m\angle I = (x + 12)°$, and $m\angle LGI = 4x°$, find $m\angle LGI$.

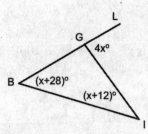

(1) $20°$

(3) $85°$

(2) $80°$

(4) $90°$

6 _____

7 Trapezoid $RSTU$ has parallel sides $\overline{RS}$ and $\overline{TU}$. Point V lies on $\overline{RS}$ such that m$\angle RUV = 58°$, m$\angle VTU = 38°$, and $\overline{UR} \cong UV$. What is the measure of $\angle TVU$?

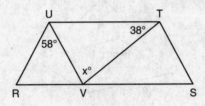

(1) 58°   (3) 74°

(2) 61°   (4) 81°   7 _____

8 If two sides of a triangle are 25 and 50 inches, which of the following could be the perimeter of the triangle?

(1) 75   (3) 115

(2) 85   (4) 150   8 _____

9 The equation of a circle is $x^2 + y^2 + 8x + 10y - 8 = 0$. What is the radius of the circle?

(1) 7   (3) 14

(2) 8   (4) 16   9 _____

10 In the figure below, $\overline{OB}$ bisects $\angle GBL$ and $\angle L \cong \angle BOG$. Which of the following statements is *not* true?

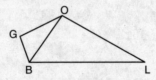

(1) $\angle BGO \cong \angle BOL$ (3) $\dfrac{BO}{OG} = \dfrac{BO}{BL}$

(2) $\dfrac{GB}{BO} = \dfrac{GO}{OL}$ (4) $\dfrac{BG}{BO} = \dfrac{BO}{BL}$ 10 _____

11 $\triangle JOG$ is a right triangle with a right angle at $\angle O$ as shown in the
figure below. If $JO = 13$ and $JG = 14.5$, what is the measure
of $\angle J$ rounded to the nearest tenth?

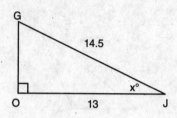

(1) 63.7°   (3) 34.7°

(2) 41.9°   (4) 26.3°   11 _____

12 Trapezoid $ABCD$ is reflected over line m as shown in the figure
below. The resulting image is $A'B'C'D'$. Which of the following is
not necessarily true about the reflection?

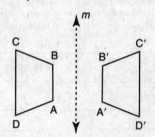

(1) Line m is $\perp$ to $\overline{AA'}$.

(2) Line m passes through the midpoint of $\overline{AA'}$.

(3) $\overline{CD} \cong \overline{C'D'}$

(4) Line m is parallel to $\overline{CD}$ and $\overline{C'D'}$.   12 _____

13 Segment $\overline{JL}$ has endpoint J with coordinates $(5,9)$. Point M is the
midpoint of $\overline{JL}$ and has coordinates $(7,6)$. What are the coordinates
of point L?

(1) (9,3)   (3) (12,15)

(2) (6,7.5)   (4) (3,12)   13 _____

14 Which statement and reason completes the proof shown below?

Given: $\angle H \cong \angle K$ and $\overline{IJ}$ bisects $\angle HIK$

Prove: $\triangle HIJ \cong \triangle KIJ$

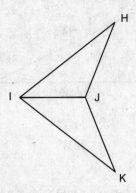

Statement	Reason
1. $\angle H \cong \angle K$ and $\overline{IJ}$ bisects $\angle HIK$	1. Given
2.	2. An angle bisector divides an angle into two congruent angles
3. $\overline{IJ} \cong \overline{IJ}$	3. Reflexive property
4. $\triangle HIJ \cong \triangle KIJ$	4.

(1) $\overline{HI} \cong \overline{IK}$, SSA

(2) $\overline{HI} \cong \overline{IK}$, SAS

(3) $\angle HIJ \cong \angle KIJ$, SAS

(4) $\angle HIJ \cong \angle KIJ$, AAS

14 _____

15 Which sequence of transformations maps pentagon $ABCDE$ to $A'B'C'D'E'$ and then to $A''B''C''D''E''$?

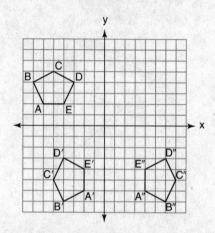

(1) a reflection over the x-axis followed by a reflection over the y-axis

(2) a counterclockwise rotation of $90°$ about the origin followed by translation of 6 units to the right

(3) a counterclockwise rotation of $90°$ about the origin followed by a reflection over the line $x = 1$

(4) a point reflection through the origin followed by a reflection over the line $x = -1$

15 _____

16 In triangle PQR, segment $\overline{MN}$ is drawn parallel to $\overline{PQ}$. Given $MN = 12$, $PQ = 14$, $PM = x - 4$, and $MR = 2x + 4$, find the value of x.

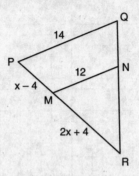

(1) $\dfrac{8}{3}$ (3) 7

(2) $\dfrac{13}{2}$ (4) 9   16 _____

17 In the circle shown below, secant $\overline{QS}$ intersects the circle at R and secant $\overline{US}$ intersects the circle at T. If $UT = 5$, $TS = 7$, and $RS = 6$, find the length of $\overline{QR}$.

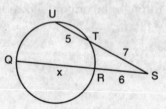

(1) $5\dfrac{5}{6}$ (3) $8\dfrac{2}{5}$

(2) 8   (4) 9   17 _____

18 In the figure below, points B, N, T, and P are collinear. If $\angle B \cong \angle P$ and $\overline{BL} \cong \overline{JP}$, which piece of additional information would be sufficient to prove $\triangle BLT \cong \triangle PJN$?

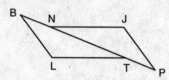

(1) $\overline{LT} \cong \overline{JP}$ (3) $\overline{LT} \cong \overline{JN}$

(2) $\overline{BL} \cong \overline{JN}$ (4) $\overline{BN} \cong \overline{PT}$ 18 _____

19 In the figure of $\triangle GHI$ below, K is the midpoint of $\overline{GH}$ and J is the midpoint of $\overline{HI}$. Which of the following statements must be true?

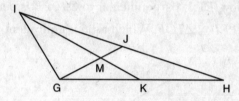

(1) $\dfrac{GM}{GJ} = \dfrac{3}{2}$ (3) $\dfrac{IM}{KM} = \dfrac{1}{2}$

(2) $\dfrac{GK}{KH} = \dfrac{1}{2}$ (4) $\dfrac{KM}{KI} = \dfrac{1}{3}$ 19 _____

20 Right triangle DJR has a right angle at D. Altitude DM is drawn to hypotenuse $\overline{JR}$. If $JM = 6$ and $DM = 9$, find the length of $\overline{RM}$.

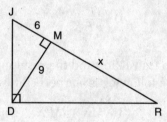

(1) 4   (3) 13.5

(2) 7.5   (4) 14.1   20 _____

21 Find the equation of the line that is parallel to $y = \frac{2}{3}x + 6$ and passes through the point (9,1).

(1) $y = \frac{2}{3}x - 5$ (3) $y = -\frac{3}{2}x + 6$

(2) $y = \frac{2}{3}x + 6$ (4) $y = -\frac{3}{2}x + \frac{29}{2}$ 21 _____

22 When the line $y = 2x + 3$ is dilated with a scale factor of 4 centered at the origin, what is the equation of the image?

(1) $y = 8x + 3$ (3) $y = 8x + 12$

(2) $y = 2x + 12$ (4) $y = \frac{1}{2}x + 3$ 22 _____

23 In the figure below, $\overline{FT}$ is perpendicular to $\overline{OT}$, $\overline{FP}$ is perpendicular to $\overline{PT}$, and $\angle PFT \cong \angle TOF$. Which ratio is equivalent to $\sin\angle OFT$?

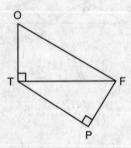

(1) $\dfrac{FP}{FT}$ (3) $\dfrac{OT}{FT}$

(2) $\dfrac{FT}{FO}$ (4) $\dfrac{TP}{FT}$ 23 _____

24 If $\triangle MNO$ undergoes a translation and the resulting image is $\triangle M'N'O'$, which of the following is *not* necessarily true?

(1) $\overline{MN} \cong \overline{NO}$ (3) $\overline{MN} \cong \overline{M'N'}$

(2) $\angle O \cong \angle O'$ (4) $\overline{MM'} \cong \overline{NN'}$ 24 _____

PART II

Answer all **7** questions in this part. Each correct answer will receive **2** credits. Clearly indicate the necessary steps, including appropriate formula substitutions, diagrams, graphs, charts, etc. Utilize the information provided for each question to determine your answer. Note that diagrams are not necessarily drawn to scale. For all questions in this part, a correct numerical answer with no work shown will receive only 1 credit.   [14 credits]

25 Directed line segment $\overline{RK}$ has the endpoints $R(1,-3)$ and $K(10,9)$.
    Find the coordinates of point P, which divides $\overline{RK}$ in a 2:1 ratio.
    The use of the grid below is optional.

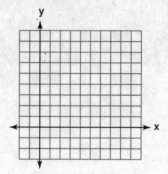

26 Triangle QVC is shown below with m$\angle V = 43°$, m$\angle C = 115°$, $QV = 16$, and $QC = 12$. Find the area of the triangle. Round to the nearest tenth.

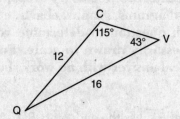

27 Parallelogram *FROG* has vertices at $F(3,1)$, $R(6,1)$, $O(5,4)$, and $G(8,4)$.

State the coordinates of $F'R'O'G'$, the image of *FROG* after a vertical stretch with a scale factor of 3.

Is $F'R'O'G'$ similar to *FROG*? Explain your reasoning.

28 Given: $\overline{WYR}$ and $\overline{TYX}$ intersect at Y

       $\overline{TR} \parallel \overline{WX}$

 Prove: $\triangle WXY \sim \triangle RTY$

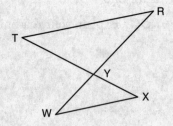

29 Rectangle $W'X'Y'Z'$ is the image of rectangle $WXYZ$ after a dilation through point P. State the coordinates of point P, and state the scale factor of the dilation.

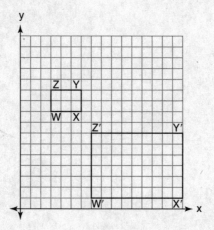

30 Circle P has a diameter of 36 feet. A central angle measuring 80°
intercepts an arc on the circle. What is the length of the intercepted arc in feet? Express your answer in terms of π.

31 *LMPR* is an isosceles trapezoid with $\overline{LR} \cong \overline{MP}$ and altitude $\overline{RZ}$. If m∠60°, $RP = 10$, and $RZ = 6$, find the perimeter of *LMPR*. Express your answer in simplest radical form.

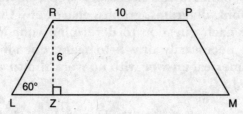

PART III

Answer all 3 questions in this part. Each correct answer will receive 4 credits. Clearly indicate the necessary steps, including appropriate formula substitutions, diagrams, graphs, charts, etc. Utilize the information provided for each question to determine your answer. Note that diagrams are not necessarily drawn to scale. For all questions in this part, a correct numerical answer with no work shown will receive only 1 credit. [12 credits]

32 Construct a circle inscribed in $\triangle FLY$ shown below. Label the center of the circle as point P. [Leave all construction marks.]

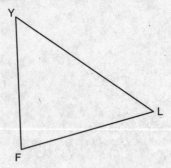

33 △*UVW* has vertices with coordinates *U*(4,8), *V*(−6,10), and
W(2,−2). Prove △*UVW* is an isosceles right triangle. The use of
the grid below is optional.

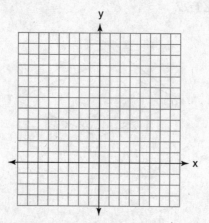

34 Given: Rhombus *MATH*

$\overline{MZ} \cong \overline{TZ}$

Prove: $\angle MHZ \cong \angle THZ$

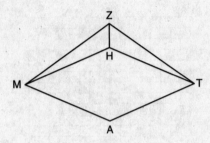

PART IV

Answer the question in this part. A correct answer will receive 6 credits. Clearly indicate the necessary steps, including appropriate formula substitutions, diagrams, graphs, charts, etc. Utilize the information provided for each question to determine your answer. Note that diagrams are not necessarily drawn to scale. A correct numerical answer with no work shown will receive only 1 credit. [6 credits]

35 A tanker truck carrying liquid propane has a tank that is comprised of a cylinder with hemispherical caps on each end. The cylindrical section is 920 cm long and 240 cm in diameter as shown in the figure below.

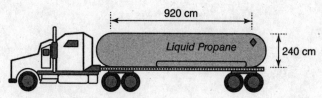

Find the volume of the propane tank. Express your answer in liters. Round to the nearest liter.

Question 35 is continued on the next page.

Question 35 continued

The density of liquid propane is 0.493 kg/liter, and it costs
$1.48/kilogram. Using your answer to the previous part, what is the
cost of a full tanker truck of propane? Round your answer to the
nearest dollar.

How many empty 300-liter residential tanks can be completely
filled using the propane from one tanker truck?

Answers Sample 2025 Exam
Geometry

Answer Key

PART I

1. 3	**5.** 1	**9.** 1	**13.** 1	**17.** 2	**21.** 1
2. 3	**6.** 2	**10.** 3	**14.** 4	**18.** 4	**22.** 2
3. 3	**7.** 4	**11.** 4	**15.** 3	**19.** 4	**23.** 1
4. 2	**8.** 3	**12.** 4	**16.** 3	**20.** 3	**24.** 1

PART II

25. $(7, 5)$

26. 36.0

27. $F'(3, 3), R'(6, 3), O'(5, 12),$ $G'(8, 12)$

They are not similar because the vertical stretch changes the shape of the figure.

28. $\angle T \cong \angle X$ and $\angle R \cong \angle W$ because alternate interior angles are congruent. Therefore, $\triangle WXY \sim \triangle RTY$ by the AA theorem.

29. $P(1, 13)$, scale factor $= 3$

30. 8π

31. $20 + 12\sqrt{3}$

PART III

32. See the construction method for an inscribed circle.

33. It is isosceles because both legs have a length of $\sqrt{104}$. It is a right triangle because the legs are perpendicular with negative reciprocal slopes of $-\frac{2}{10}$ and $\frac{10}{2}$.

34. $\overline{MH} \cong \overline{HT}$ because consecutive sides of a rhombus are congruent. $\overline{ZH} \cong \overline{ZH}$ by the reflexive property. $\triangle MHZ \cong \triangle THZ$ by the SSS theorem, which makes $\angle MHZ \cong \angle THZ$ by CPCTC.

PART IV

35. 48,858 liters
$35,649
162 tanks

In **PARTS II–IV**, you are required to show how you arrived at your answers. For sample methods of solutions, see the *Answers Explained* section.

Answers Explained

PART I

1. The plane cutting through the cylinder perpendicular to the bases is shown in the accompanying figure. The cross section is a 25×30 rectangle.

    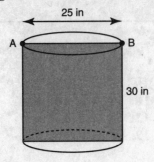

    The correct choice is **(3)**.

2. Angles 1 and 4 are alternate interior angles formed by lines p and m and transversal a. Two lines are parallel if alternate interior angles formed by a transversal are congruent.

    The correct choice is **(3)**.

3. Two pairs of parallel opposite sides are sufficient to show a quadrilateral is a parallelogram. To prove it's a rectangle, we need one of the additional rectangle properties:
    *   Congruent diagonals
    *   Sides form right angles

    The correct choice is **(3)**.

4. If we consider the segment between the preimage and image, the reflection point is the midpoint. Count jumps in both the x- and y-directions when going from the preimage to the reflection point.

    $(5, 1) \rightarrow (3, 4)$ is -2 units in the x-direction and $+3$ units in the y-direction.

Now apply these jumps to the coordinates of the reflection point to get $(3 - 2, 4 + 3)$. The coordinates of the image are $(1, 7)$.

The correct choice is **(2)**.

5. The base of the pyramid is a 27×26 rectangle. The area of the base is $27 \text{ cm} \times 26 \text{ cm} = 702 \text{ cm}^2$. Use the volume formula for a pyramid with a height of 14 cm and a base with area 702 cm^2.

$$V = \frac{1}{3}Bh$$
$$V = \frac{1}{3} \cdot 702 \cdot 14$$
$$V = 3276 \text{ cm}^3$$

The correct choice is **(1)**.

6. The exterior angles theorem states that the measure of an exterior angle of a triangle is equal to the sum of the two farthest interior angles.

$$m\angle LGI = m\angle B + m\angle I$$
$$4x = x + 28 + x + 12$$
$$4x = 2x + 40$$
$$2x = 40$$
$$x = 20$$

Substitute $x = 20$ into the expression for $\angle LGI$.

$$m\angle LGI = 4 \cdot 20 = 80°$$

The correct choice is **(2)**.

7. Triangle RUV is isosceles, so the two base angles, $\angle URV$ and $\angle UVR$, must be congruent. We can solve for the base angles using the 58° vertex angle and the angle sum theorem, which says all three angles in a triangle must sum to 180°.

$$58° + x + x = 180°$$
$$2x = 122°$$
$$x = 61°$$

We now know $\angle UVR = 61°$. We also know that the alternate interior angles formed by the two parallel sides and transversal $\overline{TV}$ must be congruent. In this case, we have $\angle UTV \cong \angle SVT$, so $m\angle SVT = 38°$.

Now we apply the fact that the sum of three angles around point V must sum to $180°$ since they combine to form a straight line.

$$m\angle UVR + m\angle TVU + m\angle SVT = 180°$$
$$61° + m\angle TVU + 38° = 180°$$
$$99° + m\angle TVU = 180°$$
$$m\angle TVU = 81°$$

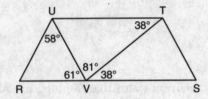

The correct choice is **(4)**.

8. The triangle inequality theorem states that given two sides of a triangle, the length of the third side must be between the sum and difference of the other two sides. In this case, the third side must be greater than $50 - 25$ and less than $50 + 25$, so the third side is greater than 25 and less than 75.

The perimeter of a triangle is the sum of all three sides. Using the minimum and maximum values of the third side, the perimeter must be greater than $25 + 50 + 25 = 100$ and less than $25 + 50 + 75 = 150$.

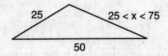

The perimeter must therefore satisfy the inequality $100 < P < 150$. The only choice within this range is 115.

The correct choice is **(3)**.

9. To find the radius of the circle, we need to rewrite the equation in center-radius form:

$$(x-h)^2 + (y-k)^2 = r^2$$

where the coordinates of the center are (h, k) and the radius is r. We can use the completing the square method to get the equation in the proper form. First group x-terms and group y-terms. Then move the constant to the other side of the equal sign.

$$x^2 + y^2 + 8x + 10y - 8 = 0$$
$$x^2 + 8x + y^2 + 10y = 8$$

Next add $\left(\frac{x\text{-coefficient}}{2}\right)^2$ and $\left(\frac{y\text{-coefficient}}{2}\right)^2$ to each side and simplify.

$$x^2 + 8x + \left(\frac{8}{2}\right)^2 + y^2 + 10y + \left(\frac{10}{2}\right)^2 = 8 + \left(\frac{8}{2}\right)^2 + \left(\frac{10}{2}\right)^2$$
$$x^2 + 8x + 16 + y^2 + 10y + 25 = 49$$

Finally factor the x-terms and the y-terms. You will get two squared binomials.

$$(x+4)^2 + (y+5)^2 = 49$$

The center of the circle is $(-4, -5)$, and the radius is $\sqrt{49}$, or 7.

The correct choice is **(1)**.

10. From the angle bisector, we know $\angle GBO \cong \angle LBO$, and $\angle L \cong \angle BOG$ is given. Therefore, $\triangle GBO \sim \triangle OBL$.

Choice (1) is true because $\angle BGO$ and $\angle BOL$ are corresponding angles in similar triangles. Choices (2) and (4) are also true because they are proportions formed by ratios of corresponding sides. Choice (3) is not true because $\frac{BO}{OG}$ represents the long side divided by the medium side in $\triangle GBO$ but $\frac{BO}{BL}$ represents the short side divided by the long side in $\triangle OBL$.

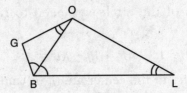

The correct choice is **(3)**.

11. Apply an inverse trigonometric ratio to find the missing angle. Relative to angle J, OJ is the adjacent and GJ is the hypotenuse. This suggests using an inverse cosine.

$$x = \cos^{-1}\frac{\text{adjacent}}{\text{hypotenuse}}$$

$$x = \cos^{-1}\frac{13}{14.5}$$

$$x = \cos^{-1}(0.896551)$$

$$x = 26.29°$$

The correct choice is **(4)**.

12. The line of reflection is always the perpendicular bisector of the segment connecting a point and its image. Line m is the perpendicular bisector of $\overline{AA'}$, which means line m is perpendicular to $\overline{AA'}$ and passes through its midpoint. Choices (1) and (2) are true. A reflection is also a rigid motion, meaning the image of a segment is congruent to the original, so $\overline{CD} \cong \overline{C'D'}$. Choice (3) is true. The line of reflection, however, does not necessarily have any particular orientation to the preimage, so we don't know for sure if m is parallel to $\overline{CD}$ or $\overline{C'D'}$.

The correct choice is **(4)**.

13. The midpoint is halfway between the two endpoints in both the x- and y-directions. Count the number of jumps in each direction from J to M. In this case, x increases by 2 and y decreases by 3. Apply that number of jumps to M to get the coordinates of L. The coordinates of L are $(9, 3)$.

$$
\begin{array}{ccc}
\bullet\!\!\!-\!\!\!-\!\!\!-\!\!\!-\!\!\!-\!\!\!-\!\!\!\bullet\!\!\!-\!\!\!-\!\!\!-\!\!\!-\!\!\!-\!\!\!-\!\!\!\bullet \\
J & M & L \\
(5, 9) & (7, 6) & (9, 3)
\end{array}
$$

$$
\begin{array}{cc}
x\!:\ +2 & x\!:\ +2 \\
y\!:\ -3 & y\!:\ -3
\end{array}
$$

The correct choice is **(1)**.

14. The angle bisector divides $\angle HIK$ into the angles $\angle HIJ$ and $\angle KIJ$, so we know $\angle HIJ \cong \angle KIJ$. Label the figure with congruent marks on those angles. Also label the given congruent angles $\angle H$ and $\angle K$, and mark the shared side $\overline{IJ}$ with congruent marks. We can see that the arrangement of sides and angles matches the AAS postulate.

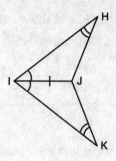

The correct choice is **(4)**.

15. A rotation of $90°$ counterclockwise around the origin maps $ABCDE$ to $A'B'C'D'E'$. The transformation to go to $A''B''C''D''E''$ is a reflection over a vertical line. To determine the line, choose a pair of points and count the distance between them. It is 6 horizontal jumps from A' to A''. Divide that distance in half and count from A' to find the location of the vertical line. Three jumps right from A' is an x-coordinate of 1, so the reflection is over the line $x = 1$.

The correct choice is **(3)**.

16. A line parallel to a side of a triangle always divides the triangle into two similar triangles. A great strategy is to split apart the two triangles and label them separately as shown, combining the two parts of $\overline{PR}$ to get a total length of $3x$. The corresponding parts form a proportion that we can use to solve for x.

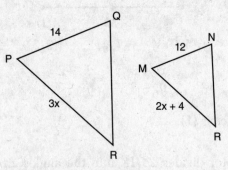

$$\frac{PQ}{MN} = \frac{PR}{MR}$$

$$\frac{14}{12} = \frac{3x}{2x + 4}$$

$$12(3x) = 14(2x + 4)$$

$$36x = 28x + 56$$

$$8x = 56$$

$$x = 7$$

The correct choice is **(3)**.

17. When a secant intersects a circle, the secant is divided into an inside part and an outside part. The relationship between the parts of two secants from the same external point is

$$\text{outside} \times \text{whole} = \text{outside} \times \text{whole}$$

By applying this relationship to the two secants, we get

$$SR \cdot SQ = ST \cdot SU$$

$$6(6 + x) = 7(12)$$

$$36 + 6x = 84$$

$$6x = 48$$

$$x = 8$$

The correct choice is **(2)**.

18. Begin by labeling congruent parts with congruent parts. Remember to label the shared part $\overline{NT}$. Note that $\overline{NT}$ is not a complete side. However, if $\overline{BN} \cong \overline{PT}$, segment addition could be used to show $\overline{BT} \cong \overline{PN}$. At that point, we would have enough information to prove the triangles congruent by SAS. None of the other choices would work. Choice (1) is not correct because $\overline{LT}$ and $\overline{JP}$ are not corresponding sides. Choice (2) also references sides that are not corresponding. Choice (3), $\overline{LT} \cong \overline{JN}$, would result in SSA, which is not a congruence postulate.

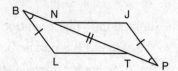

The correct choice is (4).

19. $\overline{GJ}$ and $\overline{KI}$ are medians that intersect at centroid M. A centroid divides a median in a 1 : 2 : 3 ratio, where 1 corresponds to the shortest part, 2 corresponds to the longer part, and 3 corresponds to the whole median. In choice (4), $\overline{KM}$ is the shortest part and $\overline{KI}$ is the entire median, so a ratio of $\frac{1}{3}$ is indicated. The other choices have incorrect ratios.

The correct choice is (4).

20. When an altitude in a right triangle is drawn to the hypotenuse, three similar right triangles are formed. A proportion formed by corresponding side lengths can be used to find a missing side. One strategy to determine the correct proportion is to set up a table for the three triangles—small, medium, and large triangles. Each triangle will have a short leg, a medium leg, and a hypotenuse.

	Small Triangle	Medium Triangle	Large Triangle
Short Leg	6	9	—
Medium Leg	9	x	—
Hypotenuse	—	—	$6 + x$

Look for two columns and rows that have values or variables, and the corresponding parts will line up automatically in the table. In this case, we have the small and medium triangles. Set up a proportion and solve.

$$\frac{6}{9} = \frac{9}{x}$$
$$6x = 81$$
$$x = 13.5$$

The correct choice is **(3)**.

21. Parallel lines have the same slope. The equation of a line is $y = mx + b$ where m is the slope, so the slope of the parallel line is $\frac{2}{3}$. The new line will have a different y-intercept, b. Solve for b by substituting $m = \frac{2}{3}$, $x = 9$, and $y = 1$.

$$y = mx + b$$
$$1 = 9 \cdot \frac{2}{3} + b$$
$$1 = 6 + b$$
$$b = -5$$

The new line has the equation $y = \frac{2}{3}x - 5$.

The correct choice is **(1)**.

22. When a line is dilated, the resulting image is a parallel line with the same slope and a y-intercept scaled by the scale factor of the dilation. The new line will have a slope of 2 and a y-intercept of 3×4, which is 12. The new line has the equation $y = 2x + 12$.

The correct choice is **(2)**.

23. The two triangles have two pairs of congruent angles, $\angle OTF \cong \angle FPT$ and $\angle PFT \cong \angle FOT$. The triangles are similar by the AA postulate; therefore, ratios of corresponding sides are congruent. We can write the similarity statement using a labeled figure to help us correctly match up corresponding sides, $\triangle FTO \sim \triangle TPF$.

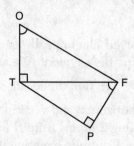

The sine ratio is equal to $\dfrac{\text{opposite}}{\text{hypotenuse}}$, so $\sin\angle OFT = \dfrac{OT}{OF}$. This isn't any of the choices, so rewrite the ratio using corresponding parts from $\triangle TPF$. $\overline{OT}$ corresponds to $\overline{FP}$, and $\overline{OF}$ corresponds to $\overline{FT}$. Applying this substitution gives $\sin\angle OFT = \dfrac{FP}{FT}$.

The correct choice is **(1)**.

24. A translation is a rigid motion, so all pairs of corresponding parts are congruent. Choices (2) and (3) are pairs of corresponding parts, so those must be true. Also, each point is translated by the same vector. The distance between any point and its image is always the same, making $\overline{MM'} \cong \overline{NN'}$, so choice (4) is true. We are not provided any information about congruent parts within triangle MNO, so we don't know if $\overline{MN} \cong \overline{NO}$ is true.

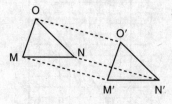

The correct choice is **(1)**.

PART II

25. This problem can be solved algebraically or graphically. For a graphical solution, begin by graphing the segment. We are looking for the location of point P that makes $\frac{RP}{PK} = \frac{2}{1}$. To find the location of P, complete the right triangle RQK by adding a horizontal and vertical segment. Then count sets of jumps horizontally, counting 2 jumps from R for every one jump from Q as shown in the figure below. Continue counting until the jumps from R and Q meet. This will be the x-coordinate. If you trace vertically to $\overline{RK}$, your line will intersect $\overline{RK}$ at P.

If you prefer, or if the horizontal jumps bypass each other without meeting, the following algebraic formula can be applied where $r_1 = 2$ and $r_2 = 1$ from the 2 : 1 ratio.

$$x = x_1 + \frac{r_1}{r_1 + r_1}(x_2 - x_1)$$

$$y = y_1 + \frac{r_1}{r_1 + r_1}(y_2 - y_1)$$

Either way, the coordinates of P are $(7, 5)$.

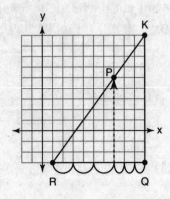

26. The formula $A = \frac{1}{2}ab \sin C$ can be used to find the area of a triangle, where a and b are any two sides and $\angle C$ is the angle formed by those two sides. From the figure, the two given sides form $\angle Q$, so we first need to find m$\angle Q$ using the angle sum theorem.

$$m\angle Q + 43° + 115° = 180°$$
$$m\angle Q + 158° = 180°$$
$$m\angle Q = 22°$$

Now apply the area formula for a triangle.

$$A = \frac{1}{2}ab \sin C$$
$$A = \frac{1}{2}(12)(16) \sin 22°$$
$$A = 35.962$$

Rounded to the nearest tenth, the area is 36.0.

27. A vertical stretch will multiply only the y-coordinates by the scale factor. The new coordinates are

$$F'(3, 3), R'(6, 3), O'(5, 12), G'(8, 12)$$

$F'R'O'G'$ is not similar to $FROG$ because a vertical stretch is not a rigid motion. The shape and angle measures of $F'R'O'G'$ are different from those in $FROG$, so the images cannot be similar.

28. The two parallel lines will form congruent alternate interior angles that can be used to show the triangles are similar by the AA theorem. Be sure to match up the corresponding pairs of angles correctly.

Statement	Reason
1. $\overline{WYR}$ and $\overline{TYX}$ intersect at Y	1. Given
2. $\overline{TR} \parallel \overline{RX}$	2. Given
3. $\angle T \cong \angle X$ and $\angle R \cong \angle W$	3. Alternate interior angles formed by parallel lines and a transversal are congruent
4. $\triangle WXY \sim \triangle RTY$	4. AA theorem

29. Find the center of the dilation by connecting pairs of corresponding points. The lines will intersect at the center point. The center of the dilation is $P(1, 13)$.

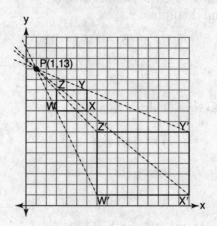

The scale factor can be found by calculating the ratio of the lengths of any pair of corresponding sides, making sure to put the image in the numerator and the preimage in the denominator. Using $\overline{WX}$ and $\overline{W'X'}$, the scale factor is 3.

$$\frac{W'X'}{WX} = \frac{9}{3} = 3$$

30. The formula for the length of an arc is

$$\text{arc length} = 2\pi r \frac{\theta}{360°}$$

where r is the radius and θ is the measure of the central angle in degrees. The radius is half the diameter, or 18 feet. Substitute the values to calculate the arc length. Since the answer must be in terms of π, do not type the π into your calculator!

$$\text{arc length} = 2\pi \cdot 18 \cdot \frac{80°}{360°} = 8\pi$$

The length of the intercepted arc is 8π feet.

31. The perimeter of a figure is the sum of all the side lengths of a figure. To get length LR in simplest radical form, we can use the 30°-60°-90° right triangle shown below. All such triangles are similar, so a ratio of corresponding parts can be used to find LR. Note that trigonometry can also be used, but the result will be in the form of a rounded decimal and not radical form. Using the 30°-60°-90° below and similar triangle LZR, form a proportion using the known side RZ and the unknown side LR.

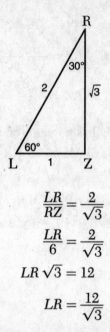

$$\frac{LR}{RZ} = \frac{2}{\sqrt{3}}$$

$$\frac{LR}{6} = \frac{2}{\sqrt{3}}$$

$$LR \sqrt{3} = 12$$

$$LR = \frac{12}{\sqrt{3}}$$

Now rationalize the denominator and simplify.

$$LR = \frac{12}{\sqrt{3}} \cdot \frac{\sqrt{3}}{\sqrt{3}}$$

$$LR = \frac{12\sqrt{3}}{3}$$

$$LR = 4\sqrt{3}$$

We will need the length of LZ as well, so repeat the process with sides RZ and LZ.

$$\frac{LZ}{RZ} = \frac{1}{\sqrt{3}}$$

$$\frac{LZ}{6} = \frac{1}{\sqrt{3}}$$

$$LZ\sqrt{3} = 6$$

$$LZ = \frac{6}{\sqrt{3}}$$

$$LZ = \frac{6}{\sqrt{3}} \cdot \frac{\sqrt{3}}{\sqrt{3}}$$

$$LZ = \frac{6\sqrt{3}}{3} = 2\sqrt{3}$$

The middle part of side $\overline{ML}$ has a length of 10, and the right and left parts both have a length of $2\sqrt{3}$, so $ML = 10 + 4\sqrt{3}$. Combine all four sides to find the perimeter of $LMPR$.

$$\text{perimeter} = LR + RP + PM + ML$$
$$\text{perimeter} = 4\sqrt{3} + 10 + 4\sqrt{3} + (10 + 4\sqrt{3})$$
$$\text{perimeter} = 20 + 12\sqrt{3}$$

PART III

32. The center of an inscribed circle is the incenter. The incenter is the point where the angle bisectors intersect. Begin by making the angle bisector construction at vertex F as follows:

    - From vertex F, make an arc that intersects the triangle at points 1 and 2.
    - From point 1 make an arc and from point 2 make an arc so that they intersect at point 3.
    - Connect F to point 3 with a straight line. This line is the angle bisector of $\angle F$.
    - Repeat at vertex L to make another angle bisector.
    - Label the point of intersection of the two angle bisectors point P. This is the incenter.

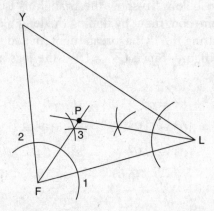

    To determine the radius of the circle, construct a perpendicular line from P to any side of the triangle as follows:

    - From point P, construct an arc that intersects the triangle at points 4 and 5.
    - From point 4 make an arc and from point 5 make an arc so that they intersect at point 6.
    - Connect point P to point 6 with a straight line.
    - Label the intersection of the line and triangle as point Q. PQ is the radius of the circle.
    - Construct a circle with the point of the compass at P and the pencil at Q. This circle is tangent to the triangle at three points and is the inscribed circle.

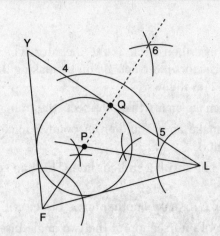

33. $\triangle UVW$ is graphed below. To show the triangle is isosceles, show that the legs $\overline{UV}$ and $\overline{UW}$ are congruent by finding the length of each. We can calculate their lengths using the Pythagorean theorem and the dashed right triangles shown in the figure. Since $\overline{UV} \cong \overline{UW}$, the legs are congruent and the triangle is isosceles.

$$a^2 + b^2 = c^2 \qquad\qquad a^2 + b^2 = c^2$$
$$2^2 + 10^2 = UV^2 \qquad 2^2 + 10^2 = UW^2$$
$$104 = UV^2 \qquad\qquad 104 = UW^2$$
$$UV = \sqrt{104} \qquad\qquad UW = \sqrt{104}$$

To show that this is a right triangle, show that the legs are perpendicular by calculating the slope of each. Perpendicular segments have negative reciprocal slopes. The slopes are negative reciprocals, so the legs are perpendicular.

$$\text{slope } \overline{UV} = \frac{\text{rise}}{\text{run}} \qquad\qquad \text{slope } \overline{UW} = \frac{\text{rise}}{\text{run}}$$

$$\text{slope } \overline{UV} = \frac{2}{-10} = -\frac{2}{10} \qquad \text{slope } \overline{UW} = \frac{-10}{-2} = \frac{10}{2}$$

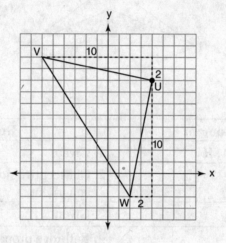

$\triangle UVW$ is an isosceles right triangle because the legs are both congruent and perpendicular.

34. First prove triangles *MHZ* and *THZ* are congruent, and then show ∠*MHZ* ≅ ∠*THZ* by CPCTC. Besides the given pair of congruent sides, rhombus *MATH* has four congruent sides. So $\overline{MH} \cong \overline{HT}$ and the shared side *HZ* is congruent to itself. All three pairs of sides are congruent, so triangles *MHZ* and *THZ* are congruent by SSS. Therefore, corresponding angles are congruent.

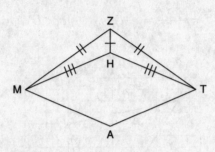

Statement	Reason
1. Rhombus *MATH*	1. Given
2. $\overline{MH} \cong \overline{HT}$	2. Consecutive sides of a rhombus are congruent
3. $\overline{MZ} \cong \overline{TZ}$	3. Given
4. $\overline{ZH} \cong \overline{ZH}$	4. Reflexive property
5. △*MHZ* ≅ △*THZ*	5. SSS theorem
6. ∠*MHZ* ≅ ∠*THZ*	6. CPCTC

PART IV

35. Find the volume of the tank by adding the volume of the cylinder to the volume of the sphere. Note that the two hemispheres combined make one full sphere. The radius of the sphere and the cylinder are the same: both are $\frac{1}{2} \cdot 240$ cm $= 120$ cm.

Cylinder	Sphere
$V = \pi r^2 h$	$V = \frac{4}{3}\pi r^3$
$V = \pi(120^2)(920)$	$V = \frac{4}{3}\pi(120^3)$
$V = 41{,}619{,}819.4747 \text{ cm}^3$	$V = 7{,}238{,}229.4738 \text{ cm}^3$

Total volume $= 41{,}619{,}819.4747 \text{ cm}^3 + 7{,}238{,}229.4738 \text{ cm}^3$

Total volume $= 48{,}858{,}048.9485 \text{ cm}^3$

Now convert to liters using the conversion shown in the reference table.

$$\frac{1 \text{ liter}}{1000 \text{ cm}^3} \cdot 48{,}858{,}048.9485 \text{ cm}^3 = 48{,}858.0489 \text{ liters}$$

Rounded to the nearest liter, the volume is 48,858 liters.

Use the density given and the calculated volume to find the mass in kg.

$$\text{mass} = \text{density} \cdot \text{volume}$$

$$\text{mass} = 0.493 \; \frac{\text{kg}}{\text{liter}} \cdot 48{,}858 \text{ liters}$$

$$\text{mass} = 24{,}086.994 \text{ kg}$$

Now apply the unit cost to find the cost of the propane in a full tank.

$$\text{cost} = \frac{\text{unit}}{\text{mass}} \cdot \text{mass}$$

$$\text{cost} = \frac{\$1.48}{\text{kg}} \cdot 24{,}086.994 \text{ kg}$$

$$\text{cost} = \$35{,}648.751$$

Rounded to the nearest dollar, the cost is $35,649.

Divide the calculated volume by 300 liters to get the number of residential tanks that can be filled.

$$\text{number of tanks} = \frac{48{,}858}{300} = 162.86$$

162 residential tanks can be completely filled. Note that we don't round up here because the tank must be completely filled.

Topic	Question Numbers	Number of Points	Your Points	Your Percentage
1. Basic Angle and Segment Relationships	2	2		
2. Angle and Segment Relationships in Triangles and Polygons	6, 7, 8	$2 + 2 + 2 = 6$		
3. Constructions	32	4		
4. Transformations	4, 12, 15, 22, 24, 27, 29	$2 + 2 + 2 + 2 + 2 + 2 + 2 = 14$		
5. Triangle Congruence	14, 18	$2 + 2 = 4$		
6. Lines, Segments, and Circles on the Coordinate Plane	9, 13, 21, 25	$2 + 2 + 2 + 2 = 8$		
7. Similarity	10, 16, 19, 20, 28	$2 + 2 + 2 + 2 + 2 = 10$		
8. Trigonometry	11, 23, 26, 31	$2 + 2 + 2 + 2 = 8$		
9. Parallelograms	3, 34	$2 + 4 = 6$		
10. Coordinate Geometry Proofs	33	4		
11. Volume and Solids	1, 5	$2 + 2 = 4$		
12. Modeling	35	6		
13. Circles	17, 30	$2 + 2 = 4$